Urban Awakenings

Samuel Alexander · Brendan Gleeson

Urban Awakenings

Disturbance and Enchantment
in the Industrial City

Samuel Alexander
Melbourne Sustainable Society Institute
University of Melbourne
Melbourne, VIC, Australia

Brendan Gleeson
Melbourne Sustainable Society Institute
University of Melbourne
Melbourne, VIC, Australia

ISBN 978-981-15-7860-1   ISBN 978-981-15-7861-8 (eBook)
https://doi.org/10.1007/978-981-15-7861-8

© The Editor(s) (if applicable) and The Author(s), under exclusive license to Springer Nature Singapore Pte Ltd. 2020
This work is subject to copyright. All rights are solely and exclusively licensed by the Publisher, whether the whole or part of the material is concerned, specifically the rights of translation, reprinting, reuse of illustrations, recitation, broadcasting, reproduction on microfilms or in any other physical way, and transmission or information storage and retrieval, electronic adaptation, computer software, or by similar or dissimilar methodology now known or hereafter developed.
The use of general descriptive names, registered names, trademarks, service marks, etc. in this publication does not imply, even in the absence of a specific statement, that such names are exempt from the relevant protective laws and regulations and therefore free for general use.
The publisher, the authors and the editors are safe to assume that the advice and information in this book are believed to be true and accurate at the date of publication. Neither the publisher nor the authors or the editors give a warranty, expressed or implied, with respect to the material contained herein or for any errors or omissions that may have been made. The publisher remains neutral with regard to jurisdictional claims in published maps and institutional affiliations.

Cover image: Jan Senbergs, Detail of *Spiral development*, 2008, acrylic on linen, 185 × 200 cm © Jan Senbergs. Courtesy Jan Senbergs and Niagara Galleries, Melbourne

This Palgrave Macmillan imprint is published by the registered company Springer Nature Singapore Pte Ltd.
The registered company address is: 152 Beach Road, #21-01/04 Gateway East, Singapore 189721, Singapore

Contents

1	A Disturbed Book: Bubbles Under the Throne	1

Part I Sleepers, Wake!

2	Unsettling the Story of Disenchantment	13
3	The Gentle Art of Urban Tramping	31

Part II BC (Before-COVID)

4	The 'New World' Is Old: Journeying Through Deep Time	47
5	Descent Pathways in a City of Gold	57
6	Adrift in the Devil's Playground	67
7	Grave Matters: Death in the Liveable City (Part I)	77
8	Cold Lazarus: Death in the Liveable City (Part II)	85

9	Walking the Corridors of Consumption	93
10	A Riverside Ramble to the Last Hotel: Lostworlders Welcome	105
11	Guardians of Gandolfo Gardens	115
12	Tramping Against Extinction: Counter-Friction to the Machine	127
13	The Monumental Army that Marches on the Spot	137
14	Sisyphus in the Suburbs: Pushing the Rock	147

Part III AC (After-COVID)

15	Virtually Tramping Through Post-Normal Times	159
16	A Time for Bad Poetry	165
17	Shimmering Text: Re-Reading *The Plague* in the Coronaverse	175
18	Care-Full Times: Suffer the Children	189
19	Rewilding the Suburbs: CERES as a Site of Enchantment	199
20	Sojourning Through a Quiet City: Envisioning a Prosperous Descent	211
21	Glitter and Doom: Between Naïve Optimism and Despair	225

Part IV Coming Through Slaughter

22	An Urban Politics of Enchantment	235

1

A Disturbed Book: Bubbles Under the Throne

Dear reader, this is not the book we planned for you. Along the way of its writing, it was blown up—into two parts—by an unforeseen and dramatic turn in the recent course of human history. You may already know what we are talking about. Yes, the global COVID-19 pandemic that surfaced first in the Chinese city of Wuhan and subsequently spread outwards with dizzying swiftness in the early months of 2020. Countless human projects and indeed human lives were suddenly extinguished by the pandemic. Much suffering and loss ensued, both in terms of lives lost and diminished by the horrid touch of the virus, as well as the epic dislocations and depredations that it imposed on social and economic life across the globe. Let us explain what this dramatic turn of events meant for the book we'd thought to present to you.

Urban Awakenings was the name we gave to the project reported in this book, which took form on Wurundjeri land. It was a project conceived in early 2019 as a series of urban investigations by the authors which set out to find novel ways of looking at the contemporary industrial city, in this case the metropolis of Melbourne where we live, capital of the State of Victoria, Australia. We were inspired

Courtesy of Pawel Kuczynski © (http://pawelkuczynski.com/).

and guided by the thesis set out in Jane Bennett's 2001 book, *The Enchantment of Modern Life*,[1] which urges a new and critical way of seeing contemporary modernity as a fractured, contradictory, unreasonable, and ultimately mortal dispensation. Far from accepting modernity's dominant narrative of disenchantment, however, Bennett seeks to tell an alter-tale, one that recognises that the world still has the capacity to enchant in ways that she maintains has ethical (and, we will argue, political) significance. We wanted to apply and extend her analysis to the urban landscape, by walking the city with eyes open to the possibility of enchantment—a methodology we describe as 'urban tramping'.[2]

Since its emergence, capitalist modernity has always attempted a ruse on humanity, because it suppresses through its arrogations of 'pure reason' the key restraining Enlightenment value of doubt. Like thinkers such as Hannah Arendt and Ulrich Beck, Jane Bennett urges us to see through the hubris of modernism and assert a radical doubt about its claims to hyperrational order, linear progress, and the ethical detachment of market rule. Her novel approach, unpacked further in forthcoming chapters, is to seek out 'enchantments' in a social dispensation that deceptively imagines itself free of this ancient value. Doing so, we argue, can unsettle the assumption that the global economic order we know today is somehow the natural modern expression of the Enlightenment. By carefully examining and experiencing the diversity of cultural forms, sites, and histories, the countless chinks that exist in capitalism's armour of self-rationalisation can be exposed.

Thus Bennett rather cheekily invites enchantment, normally an antimodern notion, back onto the agenda, not seeking to reinstate fairies, magic, or superstition, but to give licence to doubt about the claims of capitalism to be the rational, and thus, *natural* expression of modernity. Might there not be other ways to theorise and experience modernity? According to Bennett, to experience the world as merely the mechanical workings of lifeless matter, commodified and traded in a marketplace, is to see the world as disenchanted, and her (and our) concern is that the tendency of modernity to disenchant our lives has destructive social and ethical consequences. It can tempt us 'moderns' to quietly live a life of resignation, apathy, individualism, acquisitiveness, and myopia, leaving people without the necessary 'affective propulsions'[3] required to create purpose in their lives and struggle for a more humane world. A disenchanted culture is one suffering the strange ache of malaise, the cause of which is difficult to identify, like a knot of anxiety that cannot be easily untied.

To actively seek out and appreciate moments of urban enchantment, on the other hand, has ethical potential. It can give people the energy—the impulse—to care and engage, in a world that is desperately in need of ethical and political revaluation and provocation. What Bennett highlights is how the *feelings* one has participate in and shape the *thoughts* one has, and vice versa. And what people feel and think obviously affects how they act, both personally and politically. She wagers that 'to some small but irreducible extent, one must be enamoured with existence and occasionally even enchanted in the face of it in order to be capable of donating some of one's scarce mortal resources to the service of others'.[4] In this way, the interconnections between affect, thought, ethics, and politics become apparent, even if those interconnections always and everywhere remain mysterious and shifting.

We are pointing here to what might be called the affective or even aesthetic dimension of ethics and politics,[5] too often marginalised by the pose of pure reason. One cannot, we argue, even in principle, master all things in life by calculation—either physically or economically—and this critical doubt opens up theoretical space beyond calculation where moments of enchantment might be able to rewire the circuitry of the dominant imaginary and lay the foundations for alternatives to arise. We have found that meditating on and in this territory—this blurry nexus between affect, ethics, and politics—can be enlightening but also discomforting.

Indeed, Bennett begins her treatise by noting that 'a discomforting affect is often what initiates a story, a claim, a thesis'.[6] Or, in the words of political theorist John Holloway: 'The starting point of theoretical reflection is opposition, negativity, struggle.... an inarticulate mumble of discontent'.[7] In our case, a vague sense of urban disenchantment gave birth to an idea for a book, but the process of writing it (as we walked the city) somehow induced a more expansive and visionary mood, a new affective state, on account of what Bennett would call 'the wonder of minor experiences',[8] of which we will be giving account. We see and feel more in the urban landscape than we once did, and in part this book is about how and perhaps why that happened—the process. We have discovered that enchantments can disturb, and disturbances can enchant, from which we inquire: Might such affective and intellectual provocations have the potential to awaken more people from the dogmatic slumber into which our urban age has fallen? Put otherwise, can a disturbed *affect* lead to a genuinely progressive and enchanting *effect*?

Bennett wants us all to look under the shiny bonnet of the neoliberal machine to find the muffled knocks and disturbances that betray its

Courtesy of Thea T © (https://www.redbubble.com/people/tatefox/portfolio).

smooth running. This is not to question reason but rather to affirm it by resisting the structural urge of capitalism to always naturalise itself as the only feasible modern social form. Nothing in nature preordains capitalism. We know that the (really) great thinkers of the Enlightenment would support this assertion as a rational truth. They would forgive the use of enchantment to confound the cause of naturalisation.

Before we go further, a word about neoliberalism, the atmospheric political construct that whirls through and around our book. This global political project emerged in the Anglophone West during the 1970s, vowing to restore growth and prosperity to a sclerotic capitalist world economy. Geographer David Harvey explains it correctly, however, as an ideologically masked wealth grab by capitalist elites and their adherents.[9] The project has survived all attempts at political censure by weakened and increasingly divided progressive forces. Today, in an urban age, we can speak of neoliberal urbanism as a form of the project that seeks to shackle all city ambitions and interests to the cause of capital expansion and the unequal division of its spoils. With an appeal to problematic definitions of 'efficiency', this zeitgeist defers all major planning and distributive decisions to the dubious wisdom of the market.

The 'industrial city' that we set out to examine in this book has been redefined by this narrow commitment to growth by any means, at any cost. It is a city overlaid with the disenchantments of widening social division and ecocidal urbanisation, and it is a settlement that remains haunted by its colonial history. Our walking ground Melbourne is an artefact, like the many, of neoliberal urbanism, but for many millennia before us these lands were walked, and are still walked, by Australia's Indigenous peoples. We acknowledge from the outset that our project takes place on stolen Aboriginal land and that sovereignty has never been ceded. Today, as you will see, this colonial city showcases many of the trappings of the growth machine economy, including freewheeling high-rise development we term 'vertical sprawl'. This increasingly disenchanted metropolitan ground beckoned our critical attention.

* * *

So, to our project. As urbanists of liberal definition, we authors set out to apply Bennett's thesis to the question of the contemporary industrial city. We thought to reframe the project of critique she urges as a set of material—that is, *spatially enacted*—investigations of our home city Melbourne, in quest for counter-evidence to the claims of contemporary capitalism to be a natural, self-regulating order, freed from the facts of natural ecology and human frailty. We crafted an idea of ourselves as 'urban tramps' who might attempt this critical inquiry. The tramps were to be *freethinking freewalkers* of the city, encumbered only with the duty to look through the mindless objectivity of industrialism towards its troubled, contradictory soul. These irruptions of capitalist reason were to be found in the ranging, mouldering fabrics of the city.

The 'tramp' was to be an original critical identity that distinguished us, while at the same time relating us, to the various traditions of urban observation: *flâneurs*, slum journalists, ethnographers, missioners, psychogeographers, etc.[10] The crafting of this identity was partly inspired by Stephen Graham's 1926 book *The Gentle Art of Tramping*,[11] which laid out a set of desiderata for journeying that refused the dictates and assumptions of the settled life. Graham sought an unsettled life as means for self-realisation, while Bennett wanted to find disruption and contradiction in the everyday fabric of modernity.

We fused the two to create the figure of the 'urban tramp' who would seek out the discomforts and disruptions of modern life so often to be found, if one looked carefully, in the otherwise machinic confidence of the contemporary industrial city. In other words, we set out to sojourn through urban landscapes with the same sense of wonder and critical attention that a nature-walker like Henry David Thoreau embodied as he sauntered through Walden Woods. Nature can enchant and disturb, but so can the city. This book project was therefore conceived as a series of tramps around our city which would be essayed and brought together in a general published consideration that we conceived as *Urban Awakenings*. All this is explained in our early chapters to follow.

What we need to relate now—before you read another word!—is how during the working out of this project the COVID-19 pandemic changed our tramping and our publishing plans. We tell the tale simply and with confidence because the whole disruption, surprisingly (or not!), proved to be a powerful confirmation of our work and the thesis behind it. As it unfolded, the pandemic was yet another historical proof of the vulnerability and contingency of capitalism, especially its latest form, globalised neoliberalism. This huge, sudden superimposition on our project was recognised as an affirmation of its purpose.

In sharp relief our investigations, halfway through their work, encountered a city that suddenly shrugged off the long-settled diktat of neoliberalism. Australian governments at all levels were roused from their laissez-faire slumber to act swiftly and with impressive effect against the crisis. The growth machine economy was forced to a halt, as it was in much of the globe, by a state newly emboldened (or simply compelled) to protect human wellbeing. The whole time-space order of neoliberalism, including its treadmill work regimes, was suspended with great and bewildering effect.

In the contemporary parlance of children and youth, this epic drama 'broke' our project. We quickly learned that this was to good effect. The pandemic did this by imposing a new periodisation on our work as we approached publication—a separation of the investigations completed before the crisis (November 2019–February 2020) and those that proceeded after it fell significantly to our part of earth (March 2020–June 2020). We posit this new disrupted timescale as 'BC', before-COVID, and 'AC', after-COVID. And this is how the final book is framed, in two COVID parts, BC and AC (noting that AC refers to after the virus *arrived* not after it *departed*, for at the time of writing these words, still under protective lockdown, the virus remains very much on the global stage).

There is some humour and discovery in this imposed reframing. Under the State of Victoria's lockdown regulations, it was impossible to undertake the type of wide-eyed meandering that urban tramping demanded of us. Suddenly, as with many other global jurisdictions, only 'exercise' with healthful (and not, as in our case, stealthful) purpose was permitted. We complied while gently demurring with these proscriptions. A set of tramping journeys occurred after the lockdown, always with exercise as the main obvious object, and of course the tramps separated by 'safe distance'. But always with a careful eye maintained for the tramp's purpose, to find wrinkles in the self-confident fabric of neoliberal urbanism. Some virtual work was combined with real purposeful exercise as described. In short, we found our way to experience the COVID city and continue the project of seeking *Urban Awakenings*.

We know that you are reading this book with the realisation and wisdom of the many months that have followed the end of our (recorded) tramping. Much will have happened in this epic species drama since we put down our pens—even as we write the Black Lives Matter protests are erupting globally in glorious rage, suggesting that the human story is undergoing further twists and transformations, unsettling the future. Such are the timescales of publication, thankfully not much disrupted in this case by the pandemic. We have two things to say about this. First, we stand to the BC assessments we undertook of Melbourne,

1 A Disturbed Book

Courtesy of Pawel Kuczynski © (http://pawelkuczynski.com/).

our industrial city, that were made before the crisis because they were made in a city that still exists, if transformed in yet unfolding ways by the pandemic. This is the part of our scholarship that speaks to the general historical experience of neoliberalism and its urban expression in recent decades. We always wanted to interrogate the longer testimony of this city as much as its contemporary moment. Second, the COVID disruption chimes eerily in a sense with the larger thesis we carried in our journeys, that of a social order in denial about its contradictions and vulnerabilities. This realisation mid-project, and as the pandemic unfolded, added a new dimensionality to our work which we hope is captured in the AC writings.

History is always a continuously disturbed, not simply providential, process. *Urban Awakenings* testifies to that fact. In its writing, we were unexpectedly disrupted in a project that fixed on the question of disruption as a feature of the modern urban predicament. In that way the historic pandemic moment reaffirmed our undertaking. Through all this we think we have learned more deeply about the vulnerabilities of capitalist modernity than we first imagined; we see more clearly the bubbles under its throne. We offer, dear reader, a fractured book that speaks in its final form to the very question we set out with—namely, is capitalism the invulnerable, natural social form that it asserts to be? In 2020, a rogue visceral ecology, COVID-19, not an enemy empire, brought the entire system to its knees.

We consider the question answered.

Notes

1. Jane Bennett, 2001. *The Enchantment of Modern Life: Attachments, Crossings, and Ethics*. Princeton: Princeton University Press.
2. Tramping is a term used in Australia and New Zealand meaning 'bushwalking', a practice and disposition we will be applying to and in the city.
3. Bennett, *Enchantment*, p 3.
4. Ibid., p 4.
5. See Samuel Alexander, 2017. *Art Against Empire: Toward an Aesthetics of Degrowth*. Melbourne: Simplicity Institute.
6. Bennett, *Enchantment*, p 3.
7. John Holloway, 2010 (2nd edn). *Change the World Without Taking Power*. London: Pluto Press, p 1.
8. Bennett, *Enchantment*, p 3.
9. David Harvey, 2006. *A Brief History of Neoliberalism*. Oxford: Oxford University Press.
10. See, e.g., Walter Benjamin, 1997. *Charles Baudelaire*. London: Verso; Edmund White, 2015. *The Flâneur: A Stroll Through the Paradoxes of Paris*. London: Bloomsbury; and Lauren Elkin, 2017. *Flâneuse: Women Walk the City in Paris, New York, Tokyo, Venice, and London*. London: Vintage.
11. Stephen Graham, 2019 [1926]. *The Gentle Art of Tramping*. London: Bloomsbury.

Part I

Sleepers, Wake!

2

Unsettling the Story of Disenchantment

> Rising, tram, four hours in the office or factory, meal, tram, four hours of work, meal, sleep and Monday, Tuesday, Wednesday, Thursday, Friday and Saturday, according to the same rhythm – this path is easily followed most of the time. But one day the 'why' arises and everything begins in that weariness tinged with amazement. 'Begins' – this is important.[1]
> —Albert Camus

Urban life in industrial civilisation has the tendency to disenchant everyday experience. Too often, in the daily grind, one is left feeling disconnected from people, place, and purpose. We have all felt this disconnection and perhaps feel it even now—we humans of late capitalism. As if somehow aware we are fiddling while Rome (or the Arctic) burns, we might ask ourselves incredulously: What are we doing? And why? There are no clear answers to these semi-conscious disturbances. It is too easy to move through the ruts of city life with little poetry or purpose.

You can see this malaise in the slowly dying eyes of people commuting to soul-numbing jobs, those seemingly lifeless actors regurgitating the pre-written script of advanced industrial society; cogs in a vast machine, easily replaced. If one is brave enough to maintain eye contact, perhaps we see our own urban disenchantment reflected in the eyes of those tired, alienated commuters, a class into which it is so easy to fall simply by virtue of being subjects of the capitalist order. Where are we going? And why? Unfreedom persists and prevails, gazing at mobile phones, yet something in the human spirit refuses to accept that this is all there is— a vital obstinacy that helps keep despair at bay and the flicker of hope alive. The hour may be dark but on the horizon a shadow stirs. Still, the uncertain promise of some glorious new dawn is not needed to justify a rejection of a world immiserated by capital's overreach. We all know that there is more to life than *this*.

Courtesy of Lena Singla © (http://lenasingla.com/).

Unhinged from our dreams of what we hoped the world would be like, urban life today threatens us all quietly with a vague dread, a foreboding realisation that somehow the mistakes of times past have acquired inertia that is locking us into an uncomfortable existence not of our choosing.

2 Unsettling the Story of Disenchantment

We are struggling to constitute the urban future, for it seems we are constituted by the urban past. As the Anthropocene lifts its veil and we allow ourselves to digest the full extent of the social and ecological catastrophes unfolding, it becomes clear the human inhabitation of Earth is 'developing' into a story of dubious honour. Progress has begun to turn back on itself, as the promises of capital, growth, and technology fail us, despite the material benefits offered to some.

We see this with our eyes and feel it in our hearts. Most troubling of all, perhaps, we are easily left uncertain of life's meaning and direction, inducing that strange existential ache of ennui in the depths of our nature. It is a disturbing spiritual condition—an urban condition—too complex to be fully captured with words. Indeed, perhaps we would not want it fully captured with words, even if human language had the capacity. To borrow a phrase from the philosopher Ludwig Wittgenstein: 'What we cannot speak about we must pass over in silence'.[2]

And why should it be a surprise that urban life is so often disenchanting? Trees and birds are disappearing from our lives as concrete and tarseal continue their inexorable creep, exiling non-human life. We urbanites are sometimes permitted what Australians call a 'nature strip'—a small area of grass near the roadside, enclosed by concrete and beneath powerlines, upon which we can place our plastic trash trolleys each week. Our primordial essence suffocates as we lose connection with the seasons and cycles of nature, living indoors under artificial lights or in the shadow of billboards. Weeds in the pavement are doused in RoundUp in the hope of maintaining the grim tidiness of civilised life. Where is our rage for nature?

In many western cities, clone-like suburbs continue their spread into farmlands and forests. They are matched by the clone-like residential towers spreading inexorably outwards and upwards from congested inner cities. The Global South seems set to follow, voluntarily or not. In Australia, as elsewhere, urban densification has been promoted by planners and delivered by developers in quest for the 'green compact city'. It is an ideal that has been concretised, literally, by a freewheeling development industry that has produced a vast and poorly built landscape that can only be called 'vertical sprawl'. We fear this ideal to be but capitalism's latest deception, a new licence for pillage of green amenity and life space. Perhaps the compact city project is the system's final material act of violence? It is too early to tell.

When and where did things go wrong? In the midst of over-crowded cities, there is more isolation than ever, despite living on top of each other as never before. In the UK, there is now a Minister for Loneliness. If only this were some fictionalised, dystopian satire, but no, it is all too real. The pandemic has enforced through law what had earlier merely been a social fact of separation. Impatiently the cars and trucks hurry past as if they had a place to be, leaving only the smell and noise of oil's combustion in their wake. City life goes on with such fierce determination, slowed but not stopped by the coronavirus. We are easily caught up in the current, with barely enough time to breathe in the fumes or microbes. The newspapers tell of how last night there was another murder, still we casually flick through to the next article and read about sport, finance, or celebrity gossip, uncomfortably numb, anything to avoid the sinkhole of further reporting about COVID-19. If only we could see our twisted faces. If only we took time to cradle the human heart.

Still, as Albert Camus declared, one day the 'why' arises—and everything begins in that weariness tinged with amazement. 'Begins', he says '– this is important.' Weariness, Camus suggested, 'comes at the end of the acts of a mechanical life, but at the same time it inaugurates the impulse of consciousness. It awakens consciousness and provokes what follows. What follows is the gradual return into the chain or it is the definitive awakening'.[3] So which will it be? A gradual return into the chain? Or a definitive urban awakening? Surely it is up to us.

This book begins at that moment of weariness tinged with amazement. It begins when one realises that our urban vitality—our verve for city life—has begun to fade, weighed down by diverse burdens, and yet, at the same time, in a moment of madness, we capture a glimpse in almost childlike wonder of the city's prospect and lost spectacle; of what Hannah Arendt called our 'natality'—a sense that the world as it is, is not how the world has to be. New worlds—new cities—are seen partially formed between the sentences of the old story, waiting to be born. Will we live them into existence? Or will we stick our heads in the sand as the tide moves in? To resign oneself to disenchantment is to accept the latter—an orientation we set out to discredit.

2 Unsettling the Story of Disenchantment

Courtesy of Jessica Laine Morris ©
(http://jessicalainemorris.wixsite.com/artist).

An Urban Politics of Enchantment?

The premise of this book is that urban disenchantment poses an ethical and political problem. Transformative action is not set in motion merely by an intellectual appreciation of crisis, immiseration, and exploitation. One can know of these horrors and yet not act… out of disenchantment. For disenchantment's primary consequence is passive resignation to the status quo, which is capitalism's greatest achievement and its greatest tragedy. To act, to resist, to rebel, to revolt—these necessary orientations and interventions, we argue, depend on a state or mood of enchantment, the absence of which is haunting urban politics today.

Courtesy of Tomoki Hayasaka © (http://www.sheerheart.jp/).

At once, of course, the notion of 'enchantment' needs further explanation, especially in its application to the industrial city, which is arguably modernity's defining achievement. Max Weber argued that modernity was increasingly disenchanted and stamped with 'the imprint of meaninglessness'.[4] Even today the prevailing view is that the industrial city—with its cars, concrete, over-crowdedness, pollution, and noise—cannot be experienced as enchanted. Indeed, in our post-Enlightenment age, any appeal to this notion requires not just definition but justification, since it normally belongs to past ages of superstition. In this book, we seek to challenge those prevailing views, drawing on and extending the seminal work of philosopher Jane Bennett in her text, *The Enchantment of Modern Life*.[5] Inspired by Bennett's work, we will endeavour to rehabilitate the notion of enchantment and apply it to the urban context. We seek enchantment because we seek disturbance—in ways of which this book will tell. While we are the first to admit that there are plenty of aspects of contemporary life that fit the disenchantment story, we wish to test Bennett's thesis that 'there is enough evidence of everyday enchantment to warrant the telling of an alter-tale'.[6]

Let us be clear, this is no invocation to return to the oppressive superstition, stultifying tradition, and material grubbing of pre-modernity. Our values are thoroughly modern—even if we enquire, with Bruno Latour, as to whether humanity has ever been truly modern. We subscribe, that is, to the centrality of reason for human (and indeed non-human) prospect. We were reminded, however, by the late Ulrich Beck that *doubt*—the necessary restraining twin of *reason*—is also a primordial Enlightenment value. Things went very wrong in industrial modernity when doubt was cast aside in favour of the rule of excessive reason. The horrors of authoritarian and corporatist rule (Left and Right) come to mind.

This is where the notion of enchantment comes in. Our use of the word denotes not magic but the very things we might associate with healthy doubt in an industrial order; a sensitivity to ambivalence, the unresolved, the overflowing and uncontained, the surprising and unplanned for. We are open to seeing and feeling things that the cold logic of instrumental rationality might marginalise or obscure. We seek to find the rust in the machine, which may take lovely colours and remind us of the mortality of all things, of the limits to growth imposed by death and decay. Mightn't we be enchanted *with good reason* when green shoots are seen pushing through the soil as the machine of capitalism itself is composted (and composts itself)?

None of these glimpses and provocations should frighten us, as much as they might jolt or disturb us, because they work to restrain rationality and prevent the overreach reflected in assertions like limitless economic growth, unbounded abundance, and geoengineering. To look at the city anew, through the questioning lens of enchantment, is to do what therapists implore us to do through meditation: *to fall awake*.

Courtesy of Maria Peña © (http://www.maria-pena.com/).

Urban enchantment is a kind of mindfulness that is committed to awakening from the technocratic dreams and arrogations of growth fetishism and industrialism. Indeed, perhaps urbanites in the overdeveloped world may need to fall, descend, and 'degrow' from such heights in order to wake up—a complex meditation to which we will return.[7]

At base, we employ the term enchantment to signify an affective state—a mood of enchantment. We defend the idea that this mood is a necessary precondition to ethical practice and political engagement, in that it can create the emotional capacity for wonder, compassion, engagement, and generosity. As Bennett explains, to be enchanted 'is to be struck and shaken by the extraordinary that lives amid the familiar and the everyday... [it is] the uncanny feeling of being disrupted or torn out of one's default sensory-psychic-intellectual disposition'.[8] It is this surprising emotional disturbance that Bennett believes has ethical potential. To be enchanted—if only for a moment—is to see life as worth living and to see the world as a place that has the latent capacity to be transformed in more humane and ecologically sane ways. More importantly, it provides the *propulsion* to act and engage, functioning as an antidote to apathy, resignation, and perhaps even despair. An openness to enchantment might even be a form of what Jem Bendell calls 'deep adaptation'[9]—a strategy for digesting the inner dimension of the Anthropocene with courage and compassion.

It should be clear then that assessing the ethical and political potential of enchantment implies no theoretical degeneration into New Age mumbo-jumbo or any cruel aestheticism. One can never be enchanted by homelessness. But one can be enchanted by a social and political vision, and plan of action, which show why homelessness is an unnecessary feature of our cities and societies. One can never be enchanted by how the combustion of fossil fuels is drying our winters and intensifying our summers, but one can be enchanted by children going on school strikes to protest the spineless inaction of our so-called leaders. To be enchanted by 'the wonder of minor experiences'[10] helps transform the affective register of politics, by altering 'the emotions, aesthetic judgements, and dispositional moods that shape political wills, programs, affiliations, ideological commitments, and policy preference'.[11] It could be said, then, that we are exploring the political relevance of the urban mood(s).

Enchantment, in this sense, can expand the contours of what seems possible and it can provoke a revaluation of what is valued. Bennett maintains that everyday moments of enchantment can build an ethics of generosity, care, and engagement, stimulating the vital energy needed to resist injustice and participate in practices of solidarity, compassion, experimentation, and renewal. To be disenchanted is to feel one lives in a world in which meaning and purpose are absent, and in which a better world is unimaginable and so not worth fighting for. Thus disenchantment is a political and ethical problem, even as enchantment remains elusive and its experience temporary. But temporary though they are, moments of enchantment can outlive their experience, changing us forever even when the moment has passed.

We will argue that it is still possible to experience enchantment in the industrial city, despite its ugliness and violence, and in fact that this affective state is crucial to motivating the ethical and political sensibilities and behaviours needed to transform urban landscapes and trajectories. In doing so, we seek to challenge the 'narrative of disenchantment' which serves only to immobilise and demotivate collective action. The closing chapter of this book draws together insights from our perambulations and presents a sketch of what we loosely call 'an urban politics of enchantment'. This is based on a recognition that an *effective* urban politics must be an *affective* politics, one that changes (or challenges) not only how we think about the world, but also the way we feel, perceive, judge, create, and, in the end, exist in the world. As John Berger famously argued, there are different 'ways of seeing',[12] and our various urban excursions provide the building blocks for our closing statement. We seek to show that the lens of disenchantment is only one lens through which to see the modern industrial city, and a dangerous one at that, with regressive social, political, and economic implications. There are alternatives, even as we accept that the disenchanted worldview holds certain unavoidable and necessary truths. This is not a utopian or romantic book, although it retains a touch of what Terry Eagleton calls 'hope without optimism'.[13]

Thus enchantment can mean finding sensuous life in the lifeless, machinic, ultra-rationalist workings of the industrial city. It can mean seeing value beyond exchange value; worth in the worthless; and riches in ruins. In short, to find enchantment is to find a reason to live beyond capitalist reason; to look at things without the lifeless commodified gaze which values only the instrumental. If this can be done—if we can find value outside the profit-maximising rule of urban exchange value—then this sets up a wider ethics of interpretation and action outside the narrow codes, principles, and laws of the capitalistic urban process. It sets up an ethics of resistance to the rule of commodity rationality and thereby provides, or threatens to provide, a foundation for a new, post-capitalist urban politics of renewal. But we are getting ahead of ourselves.

Walking the City as the City Writes Us

Let's make clear the urban field in which we will conduct our 'action research', recorded in the following pages. Escaping the office and leaving the computer behind, our method is to walk our home city of Melbourne and be open to what it teaches; to absorb and be absorbed by the alchemy of built, social, and natural environments that constitute this metropolis and shape its cultures. As we pass through and between diverse sites of the city, we will endeavour to observe, experience, examine, and participate in our urban landscapes, cultures, and histories, attempting to distil our learnings on an eclectic range of topics that nevertheless reflect a cohesive mission: to find life and disturbance in the cracks of capitalism; to find enchantment in the industrial city. Just as Martin Heidegger once noted cryptically that humans do not speak language but rather 'language speaks man',[14] so too do we hope that the city writes us, rather than have us write the city.

That is our hope; our hypothesis. In this book, we will be reporting and reflecting on excursions that take place within Melbourne, a widely spread industrial city of about five million souls, and countless non-humans, situated at the southern base of the Australian continent. Encompassing nearly 10,000 sq kms, and wrapping to the north, east, and increasingly the west of the beautiful, if tempestuous, Port Phillip Bay, the city at the time of our journeys is one of the fastest growing in the (over)developed world. It is a 'new world' city, like those in North America and New Zealand that were established in the wake of European invasions of ancient settled lands, prior to and during the intellectual, technical, and economic revolutions that produced industrial capitalism.

Founded in 1835, Melbourne was one of the more recent additions to the 'new world' city order. It grew quickly, however, and by the latter decades of the nineteenth century was one of the largest cities in the British Empire. During his visit to the city in 1885, the influential English journalist George Augustus Henry Sala coined the phrase 'Marvellous Melbourne', which justifiably described a thriving, beautiful metropolis. But of course, as with all things in capitalism, the rule of finitude was reimposed brutally by a profound economic crash just after Sala's tribute was made. Like all industrial cities, Melbourne has experienced regular cycles of boom and bust. Presently, its rapid demographic and economic growth is overshadowed by the spectres of climate change, social polarisation, worsening homelessness, pollution, resource depletion—and, most recently, pandemic. It's harder and harder to describe this injured but still beautiful city as 'Marvellous'.

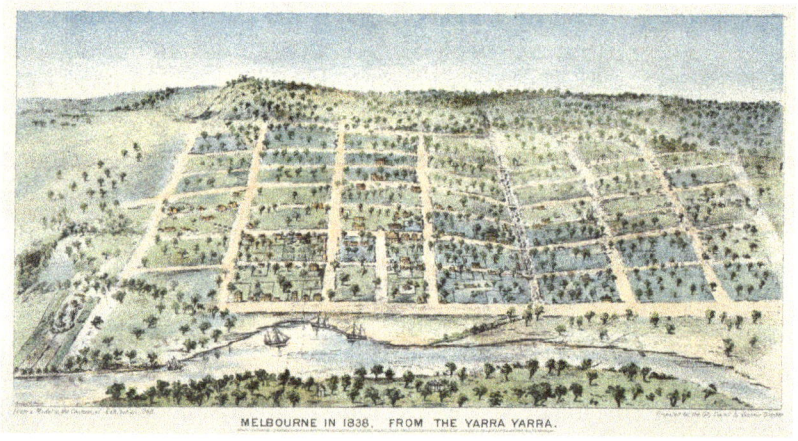

Clarence Woodhouse, Courtesy of the State Library of Victoria

Marvellously (with irony) Melbourne has no government—a degree of control over its life and shaping is exercised by the State of Victoria, of which it is the capital, and the thirty-one municipalities that constitute the metropolitan area. It is a city that has grown convulsively in recent times, adding more than 500,000 souls to its population in the last five years. The Golden Idol of Growth is worshipped in political discourse, though civil society remains somewhat less convinced. The various mainstream and social media are replete with social anxiety about the pace of growth and its many outfalls.

Contemporary corporate spin would have it otherwise. The long history of city boosting has been supplemented in recent times by the rhetoric of 'liveability'. The oxymoronic sounding 'Economist Intelligence Unit' has produced annual liveable city leagues and rankings that bestow favour on the fortunate and the canny (best when you have

Courtesy of Helen Lamb ©

both). Currently, Vienna wears the crown but it was Melbourne's prized possession between 2011–2017. Now we snake just behind Vienna, a close second. Given what we said above, the liveable city rankings (of the Economist Intelligence Unit and several other makers) must be treated with great caution. The cruelly ambiguous liveable city trope is something to think about for those wishing to take issue with the disenchantments of contemporary industrial capitalism. As we will do.

So much for modern Melbourne, our field of adventure and exploration is also an ancient settlement, the lands of the Wurundjeri, Boonwurrung, and Wathaurong peoples who had thrived within its bounds and on its bounty for millennia. It is a matter of record and extreme injustice that these people were very harshly, often barbarically, treated by settler invasions, which included a huge human influx during the nearby gold rushes of the 1850s and continued in regular cycles to today. Despite a series of massacres, dispossessions, and debauchments, the Indigenous peoples survived. They strive today, with their supporters, for the recognition and repossession of the country and culture that are their rightful inheritances. To succeed, these strivings may invite forms of enlightened enchantment in the city, acts of pride and reinstatement that refuse the narrative of a modern human settlement that somehow magically appeared in 1835.

Given its deep historical timescape, the pre-history of Melbourne must be considered the real history of the land upon which our searching occurs. These prehistories and their surviving contemporary legacies are riches stored in the city's varied landscapes: its inner-city towerscapes, ringed by graceful if increasingly congested Victorian suburbs that give way to the suburban tracts that roared into life during the twentieth century and which continue to push outwards today.

For decades, it has been a commonplace to describe most European and new world cities including Melbourne as 'post-industrial', denoting the mass exodus of manufacturing, largely to the Global South, that began with vigour from the 1970s. For Melbourne this continues to the present day, with its recent loss of a considerable automotive manufacturing sector to the vicissitudes of the global neoliberal economy. Post-industrial cities are held to be dominated by their service sectors, with many dreams about their new potentialities spun in the fine threads of the 'knowledge city', the 'smart city', and the 'green city'.

So-called post-industrial cities are made to wear all these layering garments betokening new hopes and new awakenings in the afterwashes of (often painful) deindustrialisation. Yet, we assert that underneath still lies an industrial order that is too hastily discarded or glossed over by the dreamy-eyed celebrants of post-industrialism. One must not forget that 'service' and 'information' economies inevitably rely on material and energy foundations, and evidence suggests that for all the hype, these supposedly dematerialised or decarbonised urban oases of 'green growth' have by and large turned out to be mirages.[15] The post-industrial urbanites and the system within which they produce and consume are as impactful as ever.[16]

Put simply, the post-industrial city is a dependent product of the globalised industrial order—and thus not post-industrial at all. Take away the fossil-fuelled imports, cars, trucks, technologies, and cultures, and we'll see then what a post-industrial city looks like. One way or another, for better or for worse, that future is on its way, due to the depletion of our finite inheritance of fossil fuels or the disruptive transition to renewable energy technologies for climate mitigation (or both).[17] But for now, let us not pretend otherwise: we inhabit an industrial city still trying to rebirth itself. Readers may reflect for themselves on whether their own cities have been marketed falsely as post-industrial.

Furthermore, we think expansively, and we believe correctly, about the term industrial, which at its root means much more than manufacture. Its historical form simply denotes human industry, the steady and habitual effort that is required each day to reproduce life. In this sense, Melbourne and all such cities are cities of industry, recognising also the natural processes that could equally be described as industrious. And words expand and morph over time. A more recent use of industry connotes scale, meaning that any large process, not just manufacturing, can be considered industrial.

Contemporary Melbourne like many cities betrays in numerous ways this contemporary meaning of the term, from its rampant population growth, to mounting congestion, its new monster infrastructure, and the vertical sprawl of its high-density developments, smoothing out to the myriad ordinary ways in which daily life is reproduced in the theatres and circuits of commerce, education, government, transport, and even the recreation enjoyed in its fields and stadia. The final affirmation of the term, suspended over the metropolis, is the dominant ideology of limitless growth. An iron compact of business and political elites that continues to prescribe and prosecute this value in the face of mounting evidence of its failures and deepening social mistrust. The putative commitment to compact city planning, shared with most western cities, is no barrier to growth that now occurs as much vertically and through the intensification of everyday life. In this sense, the post-industrial city is a myth; the first spell from which we must awake in seeking enchantment in the industrial city.

In these pages, we will find it in Melbourne, a city of industry; of generally peaceful industrious people but increasingly troubled and injured by the excessive reason of accumulative capitalism. There is nothing post-industrial about a metropolis, like many others, increasingly subjected to the rule of the algorithm. The so-called platform economy—dominated and defined by the digital matchmakers, Uber, Airbnb, Google, social media, and the like—is the augury of a new industrial age that demands our doubtful, critical engagement. Sarah Barns speaks of the rise in cities of 'platform urbanism'[18]—a new and imminent machinic transformation of urban functioning and life which beckons the disruptive work of the seekers of enchantment. Can we find rust already in the iron workings of their algorithms?

Our contribution in this book is to explore disruptions, cracks, and alleyways in the narrative and testimony of industrial urban Melbourne. We will report and reflect on our critical wanderings through and upon various urban and suburban landscapes, as we open our eyes to the possibility of enchantment in the spectacular diversity of spaces in the industrial city, from the grand, beautiful, prominent, and imposing, to the hidden gems and disturbances that lie beyond the gaze of tourists and glossy magazines. Nevertheless, we offer no pretensions about this being a comprehensive critical analysis or complete review of Melbourne's urban landscapes. It is but a passing and partial intervention. All the same, we hope that our Melbourne journeys and provocations will find wider resonances in a global urban age increasingly in the thrall of machinic and cybernetic thinking. We hope to rediscover a different, more marvellously enchanted Melbourne—a new Atlantis beneath our feet—by letting our senses wander with our thoughts. For what exists in the city is not what one looks at, possesses, or owns, but what one sees and feels.

Notes

1. Albert Camus, 2000 [1942]. *The Myth of Sisyphus*. London: Penguin, p 19.
2. Ludwig Wittgenstein, 1961. *Tractatus Logico-Philosophicus*. London: Routledge and Kegan Paul, p 151.
3. Camus, *Myth*, p 19.
4. Max Weber, 1981. 'Science as a Vocation' in Weber, *From Max Weber: Essays in Sociology*. Oxford: Oxford University Press, p 140.
5. Jane Bennett, 2001. *The Enchantment of Modern Life: Attachments, Crossings, and Ethics*. Princeton: Princeton University Press.
6. Ibid., p 4.
7. See also, Samuel Alexander and Brendan Gleeson, 2019. *Degrowth in the Suburbs: A Radical Urban Imaginary*. Singapore: Palgrave.
8. Bennett, *Enchantment*, pp 4–5.
9. Jem Bendell, 2018. 'Deep Adaptation: A Map for Navigating Climate Tragedy' (IFLAS Occasional Paper 2).
10. Bennett, *Enchantment*, p 3.
11. Jane Bennett, 2002. *Thoreau's Nature: Ethics, Politics, and the Wild*. Lanham: Rowman & Littlefield, p xxii.
12. John Berger, 1990. *Ways of Seeing*. London: Penguin.
13. Terry Eagleton, 2017. *Hope Without Optimism*. New Haven: Yale University Press.
14. As referenced in Richard Rorty, 1989. *Contingency, Irony, and Solidarity*. Cambridge: Cambridge University Press, p 113.
15. See generally, Jason Hickel and Giorgos Kallis, 2019. 'Is Green Growth Possible?' *New Political Economy*. https://doi.org/10.1080/13563467.2019.1598964.
16. Blair Fix, 2019. 'Dematerialization Through Services: Evaluating the Evidence'. *Biophysical Economics and Resource Quality* 6(6). https://doi.org/10.1007/s41247-019-0054-y.
17. Samuel Alexander and Joshua Floyd, 2019. *Carbon Civilisation and the Energy Descent Future: Life Beyond this Brief Anomaly*. Melbourne: Simplicity Institute; Joshua Floyd et al., 2020. 'Energy Descent as a Post-Carbon Transition Scenario: How "Knowledge Humility" Reshapes Energy Futures for Post-Normal Times'. *Futures* (in press).
18. Sarah Barns, 2020. *Platform Urbanism: Negotiating Platform Ecosystems in Connected Cities*. London: Palgrave Macmillan.

3

The Gentle Art of Urban Tramping

In 1926, the British writer-sojourner Stephen Graham presented a lyrical little text that described *The Gentle Art of Tramping*.[1] It logs wisdom and witticisms drawn from his own long journeys through Europe, including the vast and wilder territories of Russia. The book lays out a manifesto for the wanderers of the modern world, tramps not hobos, who must frown upon this world's dissolution and dissembling (with some class overtones it must be said). The tramp is a spirited 'seeker' who will not have a bit of this modern hubris and has left behind the trappings and diktats of the settled world in quest for enlightenment and the pleasures that flow from this.

Tramping, Graham explains, is 'the grammar of living',[2] a 'rebellion against housekeeping and daily and monthly accounts',[3] motivated by curiosity, wonder, and adventure, not economic imperative. 'In tramping you are not earning a living, but earning a happiness'.[4] One does not need to be rich to walk, of course—the act is free and can be freeing. It occurs 'beyond the market', constituting a post-capitalist form of journeying through time, space, and place, dependent only on the most primitive of technologies: one's legs.

Courtesy of Jorge Alaminos Fernández © (http://litoralgrafico.tumblr.com/).

Though tinged with a romantic sheen, Graham's ode to tramping is freighted with the serious idea that everyday modern life conspires remorselessly to stultify human sensibility and insight. It is a mindless (if not purposeless) living regime that reinforces what may not be ultimately good for us, a morbid economic order that brooks no doubt. We are told there is no alternative. To borrow the more recent memorable words of Slavoj Žižek, it acts to 'normalise the apocalypse'.[5] The tramp is the 'gentle rebel' who refuses this stultification by seeking out the abnormal, eyes wide open to the disquieting phenomena that lurk within all settled and wild landscapes. We adopt Graham's notion of tramping as a guiding idea for our own quest for enlightening disturbance in the industrial city. His tramp is a starting figure that we extend to embrace Bennett's more arched interrogation of capitalist modernity.

3 The Gentle Art of Urban Tramping

We recognise that our tramping notion rubs against the cultural grain in the Antipodes from where we write. It is an unsettlement in itself. In Australia, tramping might betoken heavy-booted, not gentle, behaviour. In nearby New Zealand, bushwalking or hiking is referred to as tramping. It signifies the practice of walking through 'the wild', a context that by conventional definition is distinguished from the city. Indeed, most bushwalkers would probably explain their love of tramping precisely because it is a welcome relief or an escape from the intense, hyper-organised hustle-and-bustle of city life.

And what urbanite has not breathed more freely once liberated from the big smoke? We all know that walking through the 'uncivilised' bush can be a mysterious pleasure, absent of cars, timetables, televisions, advertising, crowds, and the often grating sounds of civilisation. The hypnotic audio atmosphere of birds, insects, and softly swaying trees remind us of their industriousness, leaving us to reflect uncomfortably on our own. Free from concrete and tarseal, the moss and mycelium beneath the bushwalker's feet hide an Internet of biological connections and networks, almost as old as time itself. Interbeing is everywhere and obvious.

The affective, quasi-spiritual effects of bush hiking deserve brief elaboration. Tramping through the bush, as 'part and parcel of nature',[6] one can be alone but not necessarily lonely. It can be a special delight when traversing wild landscapes to discover that every direction looks the same, for this can give the impression that space has imploded into a simpler, single dimension, where not even time complicates things. Sauntering deeper into the woods, the city now a distant, fading memory, one's primordial being is lifted to the surface. It means finding oneself stepping more lightly, shedding the vague worries and anxieties that burden an urban existence, leaving them behind, as compost, to rejuvenate the earth that rejuvenates us. Nothing quite compares to a bushwalk at dawn, rising with the sun and awakening with each step as the communities of life (which never sleep) emit the sounds of silence, as if to greet us. These wild, almost mystical experiences can shape the soul in ways that cannot be undone. Nature is a rich reservoir of enchantment.

Nevertheless, why should we not walk through the city with the same openness to mystery, wonder, and wildness that the pioneering American environmentalist Henry David Thoreau embraced as he walked

Courtesy of Maria Peña © (http://www.maria-pena.com/).

through the woods surrounding Walden Pond? In fact, even Thoreau—who felt the need to saunter through the woods for several hours every day—closes the opening paragraph of his masterpiece *Walden* with the admission that, upon returning to his hometown of Concord after two years living in the woods, he became a 'sojourner in civilised life'.[7] That is, even as he left the wildness of Walden Woods, his life journey continued in the same spirit of learning, curiosity, and intrigue, even if his words carried with them a hint of sadness; of wildness contained and yet uncontainable.

Although we, your authors, have not had the good fortune of spending the last two years living in the woods, we look to Thoreau, Graham, and the freed spirits of the industrial world as an inspiration as we sojourn in civilised life, as urbanites, in search of the disturbance, insight, and enchantment that arises from watchful walking. Thoreau is a fitting philosopher to help ground this book, not only because he presented an early critique of industrialisation in *Walden*, but also because of his celebration of 'walking' in his famous essay by that name. Thoreau says that true Walkers are a rare breed, and that, in fact, not many people really know how to walk—exemplifying his dry and often acidic wit. We must walk like camels, he contends, 'which is said to be the only beast which ruminates when walking'.[8]

Courtesy of Robbi Wymer © (http://www.robbiwymer.com.au/).

If stepping through nature has the potential to awaken the spirit, perhaps the wildness of urban life might be able to do the same, if we let it? After all, should not the nature-city distinction itself be deconstructed to remind us that the city is and will always be part and parcel of nature? In the city too, interbeing is everywhere, if not always obvious. And Bennett is the first to admit that, while nature can certainly enchant, so do artefacts. The inventions, arts, products, and cultural complexities of humankind can also 'provoke wonder, surprise, and disorientation'.[9] That is certainly true of the city and in the city.

Can reframing the city induce enchantment? Or equally, can flashes of enchantment jolt us into a reframing of the city, broadening our channels of urban understanding and exposing blind fields in our worldviews? The point is to disrupt the normalising claims of everyday appearances and to disrupt the industrial 'settled order' that is coaxing us towards planetary overshoot and destruction. The same alleyway, park, or street lamp in the city might be received by our consciousness very differently depending on the attitudes one brings to experience—and our attitudes, as the Stoics argued, are within our control, if only we choose to shape them, to sculpt them, like a work of art. As we search for moments of urban awakening, in order to see the city and its prospects with fresh eyes, this book reports on the paradoxes, pleasures, and pitfalls of our urban tramps, or rather, the urban tramps are reported through us.

Fundamentally, then, this book is a product of 'urban tramping as method'. It draws inspiration and content from the variety of urban tramps we undertook over the last year or so, as we moved, journeyed, observed, discovered, wandered (and wondered), got lost, found ourselves, listened, smelled, touched, and felt our way through the main streets and back alleyways, the CBDs and suburbs, the parks, cemeteries, and industrial centres, the buildings and the cultures, and the peaks and troughs, of our home city of Melbourne. We were open both to the extraordinary and to what urbanist Georges Perec would call the *infraordinary*[10]: the unexpected beauty of what happens when nothing 'special' is happening.

In being self-appointed 'observers of urban life', we need to acknowledge, while also distinguishing ourselves from, the other ways in which revelation has been sought through urban journeying. Most notable is the much admired and emulated project of the French *flâneur*, which enjoined its adherents to wander without attachment the bustling streets and arcades of the early industrial city and to absorb the urban spectacle. Charles Baudelaire was its most famous exponent. Ostensibly, the *flâneur* walks blindly but in fact, as Walter Benjamin explained, secrets '…the gaze of the alienated man'.[11]

3 The Gentle Art of Urban Tramping

Courtesy of John Holcroft © (http://www.johnholcroft.com/).

To practise alienation from the vicissitudes of the everyday is the first step of our tramping but falls short of its radical striving to go further and find the fugitive logics imprisoned by the iron rationality of market society. As for the *flâneur*, feminist insight from Lauren Elkin questions the claim to alienation of a historically male sensibility, freed from the gendered circumscriptions of the industrial city.[12] This and so many other curbs are to be acknowledged in urban journeying. Your tramps are all too aware that they must walk gently on contested human grounds, most often scarred with the arrogations and injustices of power, to which first attention must be paid. Whether it is day or night, we all walk in the shadows of (usually) male violence, and this shapes everyone's experience of the city, women's especially. And in Melbourne, as in all 'new world' cities, we must acknowledge the greatest and most violent of urban alienations, which was and continues as the expropriation of Indigenous people's lands.

We recognise without hope of properly honouring in these pages the myriad other works that have opened a sceptical eye on the industrial city and its imprisoning, limiting rationalities. More politically-minded descendants of the Baudelairian *flâneur* were the 'Situationists' who operated by *dérive*, or 'drifting', offering a critique of post-war urbanism and mapping the emotive forces operating within the city and assessing their political significance.[13] In a different vein again, the late W. G. Sebald produced the magisterial tour *Rings of Saturn: an English Pilgrimage*[14] that reminded us of the terrible historical twin, fixity and decay, that corrodes the modern claim to progress. In our own Melbourne, Sophie Cunningham has looked through its hardscapes to wander and ponder *A City of Trees*.[15] Her stories belie a city celebrated internationally as 'liveable' but which is in fact progressively losing tree coverage and thus life force to freewheeling growth, profit, and intensification.

By way of fleeting review, we would be remiss not to mention also the inimitable musings of Will Self, in books like *Psychogeography*,[16] where he walks through the vortexes and fields of the city to discern the rhythms and tempos of urban life, restoring its untamed and unexpected qualities. Amongst other walking writers and philosophers, we should also note Rebecca Solnit (*Wanderlust*)[17] and Iain Sinclair (*Ghost Milk*),[18] the latter aptly described as a book in which 'an urban wanderer, local historian, avant-garde activist, and political polemicist meet and coalesce'.[19] The territory is rich.

We had this broad company of urban pilgrims in mind when making our own journeys, which now inform the critical 'journal entries' of this book. Some of these excursions we took together; others were reveries of the solitary walker. And beyond the confines of this specific project we have, of course, wandered the city with other urban tramps, meeting (or bidding farewell) to companions or each other in the midst of life's urban journeys. This extended sociality helped to expand the sensibilities and perspectives of your middle-aged, white, male authors. The city, we know very well, is a pluriverse of experience and understanding, which no author or body of authors could ever hope to fully capture or contain. Our perspective is inevitably limited, biased, and incomplete.

And so, while only two authors are listed on the cover, we wish to acknowledge that this book, like most books, is the result not merely of writers sitting at their desks drawing on their personal, self-contained 'repository of wisdom', but is a product of finding words in the blurry collective resources of shared experiences, conversations, books, spaces, and engagements which together lead to that surrealist experience of 'coming up with a sentence'. Where did these words come from? Was the idea formed when the sentence began? Or can we at times arrive at ideas or insights at the end of a sentence (or city street) that we did not know we knew when we began writing (or walking) it? More disconcertingly, who is the 'self' that typed this sentence or walked this footpath? And to what extent is the self shaped by, defined in relation to, and therefore part of, all that is 'other'? We set out disturbed and a little enchanted by our method.

Nevertheless, we choose not to tramp too close to post-structuralism and deconstruction, for risk of degenerating into unnecessary obscurantism. But one can learn things from these interpretive approaches when observing from a safe distance. Our point, for present purposes, is simply that the fragmented, decentred authorial 'we' used in this book is, first, an inconsistent and changeable alchemy of our different authorial perspectives, but more broadly, we have been inevitably and gratefully influenced by the range of interactions (with people, places, words, and things) that have shaped 'us' throughout our lives and in particular over the course of the urban wanderings we are soon to recount.

Despite these variables, what was constant in our methodology of urban tramping was the search, not so much for knowledge, but for experience upon which insight could or might arise when out walking the city and the suburbs. Sometimes silence reigned on these urban tramps as we found ourselves in thrall to the meditative process of stepping through the landscapes slowly and with deliberation; other times there was a constant dialogue with each other, or a reflective and productive monologue with oneself, as one attempted to chew through and digest the thoughts that presented themselves (uninvited) in response to our slowly shifting urban contexts.

Thus, rather than practising the conventional academic method of crafting or delivering scholarly contributions from the study or the lecture theatre, we have instead practised *urban tramping as method*. The city has been our laboratory. This book is our results and discussion.

And just to avoid giving rise to any false expectations, any conclusions we offer are suggestive and exploratory, intended only to provoke rather than compel new ethical and political directions and sensibilities, in due acknowledgement of the ever-shifting sands beneath our feet. Who wants dogmatic conclusions when urban life is an ever-changing process? As Heraclitus might have said if he were alive today: one never steps into the same city twice. Our before-COVID and after-COVID experience of the city testifies to this rather starkly. It follows that urban conclusions are always out of date. Instead, we stand by our starting point: ever industrialising capitalism is no ordained reality; it is in fact a dangerous spell from which we must awake.

What, then, did we hope to learn or gain from urban tramping? The question, as we have implied, is mis-framed, since it suggests we had a predetermined or preconceived goal. If we were seeking enchantment in the industrial city, we didn't yet know what this might mean in practice. Could this practice lead us to places that defied reason—to previously unthinkable places that expanded the imagination or that ought not to exist—thereby harbouring fugitive logics that evoked different ways of seeing, feeling, and being? Would these urban pilgrimages be joyful, disturbing, enlightening, depressing, uplifting, therapeutic, or by turns all of these things and more? Could we find moments or places of enchantment even in an industrial city that we knew marginalised the suffering and pain of both people and planet while also offering rich and diverse cultural experiences? How can these different faces of the city be reconciled?

Sometimes we would tramp with purpose and direction, having an understanding of where we were going and why, even if we discovered that our preconceptions delimited the range of experience offered as our path evolved from plan or vision to action and reality. But just as often we wandered with deliberate directionlessness and abandon, in order to receive what the city felt like giving up to the speculative eye. As J. R. R. Tolkien once wrote: 'Not all those who wander are lost'.[20]

It is time to start tramping; let us just say one more thing by way of introduction. We began with a guiding assumption—a hypothesis to test—namely, that the industrial city is 'sprinkled with natural and cultural sites that have the power to enchant'[21] and which can challenge the image of modernity as completely disenchanted. For Bennett, our philosopher of enchantment, the question is not 'whether disenchantment is a regrettable or progressive historical development. It is, rather, whether the very characterization of the world as disenchanted ignores and then discourages affective attachment to the world'.[22]

The question is important, she notes, 'because the mood of enchantment may be important to the ethical life',[23] and a post-capitalist ethics can mean an engagement with and in the world for reasons other than purely rational, self-interested, utility maximisation. Ethical progress implies, as philosopher Wilfrid Sellars argued, an expansion of 'we consciousness', bringing an ever-widening class of people, places, and wildlife within an expanding sphere of solidarity and care. Bennett argues that 'the contemporary world retains the power to enchant humans and that humans can cultivate themselves so as to experience more of that effect'.[24] In writing this book, we set out to test Bennett's thesis, but we feel we ended up performing it.

The form and style of a book can both enable and constrain what can be conveyed substantively. It follows that unconventional theses and perspectives may need to be expressed in unconventional ways. In this book, we explore the urban condition of our industrial civilisation as we tramp through our home city of Melbourne. We trust that this unusual method will allow us to say things that could not have been said any other way.

But enough words here and now. We must practise the gentle art of urban tramping. This means we must dream a little to shake away the heavyset everyday normality of the industrial city as we all find it and to liberate ourselves from the tacit assumptions of everyday life. There is another subordinated reality to be found, which quietly shouts against the looming apocalypse. Stephen Graham in 1926 sensed it. He tells his tramps, after Richter, that 'We are near awakening when we dream that we dream'.[25] In the age of Trump and the oligarchs, and the radical reassertion of capitalism's suicidal drive, we set out to denormalise the apocalypse. With a nod to Johann Sebastian Bach, *Sleepers, Wake!*

Courtesy of Jorge Alaminos Fernández © (http://litoralgrafico.tumblr.com/).

Notes

1. Stephen Graham, 2019 [1926]. *The Gentle Art of Tramping*. London: Bloomsbury.
2. Ibid., p 31.
3. Ibid., p 26.
4. Ibid.
5. Slavoj Žižek, 2010. *Living in the End Times*. London: Verso.
6. Henry Thoreau, 'Walking', in Carl Bode (ed.), 1982. *The Portable Thoreau*. New York: Penguin, p 592.
7. Henry Thoreau, *Walden*, in Carl Bode (ed.), 1982. *The Portable Thoreau*, p 258.
8. Thoreau, 'Walking', p 596.
9. Jane Bennett, 2001. *The Enchantment of Modern Life: Attachments, Crossings, and Ethics*. Princeton: Princeton University Press, p 171.

10. See Lauren Elkin, 2017. *Flâneuse: Women Walk the City in Paris, New York, Tokyo, Venice, and London*. London: Vintage, p 5.
11. Walter Benjamin, 1997. *Charles Baudelaire*. London: Verso, p 170.
12. Elkin, 2017. *Flâneuse*.
13. See Guy Debord, 1956. 'Theory of the Derive'. Available at: https://www.cddc.vt.edu/sionline/si/theory.html (accessed 5 June 2020).
14. Winfried Georg Sebald, 2002. *The Rings of Saturn*. London: Vintage.
15. Sophie Cunningham, 2019. *City of Trees: Essays on Life, Death, and the Need for a Forest*. Melbourne: Text Publishing.
16. Will Self, 2007. *Psychogeography*. London: Bloomsbury.
17. Rebecca Solnit, 2000. *Wanderlust: A History of Walking*. London: Penguin.
18. Iain Sinclair, 2011. *Ghost Milk: Recent Adventures Among the Future Ruins of London on the Eve of the Olympics*. New York: Faber and Faber.
19. Merlin Coverly, 2010. *Psychogeography*. Harpenden, UK: Pocket Essentials, p 122.
20. J.R.R. Tolkein, 2008 [1954], from the poem 'All That Is Gold Does Not Glitter' in *Fellowship of the Ring*. London: HarperCollins.
21. Bennett, *Enchantment*, p 3.
22. Ibid.
23. Ibid.
24. Ibid., p 4.
25. See Graham, *Gentle Art*, p. 152.

Part II

BC (Before-COVID)

4

The 'New World' Is Old: Journeying Through Deep Time

We, the children of industrial civilisation, do not know how young we are. Not even newborns, we are but cultural embryos, destined, it seems, to miscarry—for better or for worse. But as with all things in this finite biosphere of natural cycles and seasons, death is merely the precursor to new life, new beginnings, which in turn will end as something new emerges, over and over again, for eternity. History is this process, this flow, repeatedly lapping against the shores of deep time. From a geological perspective, the Anthropocene is but a blink of the eye, not so much an epoch as an event. It too will pass and one day be a memory, hard though that is to imagine today. For now, however, most human beings live in urban contexts that are both the product and process of industrialisation. Our species has become *homo urbanis*.

Leaving Melbourne University's Parkville Campus at the end of the workday, we meander down one of the main CBD drags, Elizabeth Street, in the direction of Flinders Street Station—towards the beating heart of the city and a defining artefact and enabler of the industrial order. It seems like a suitable landmark and symbol to inspire our inaugural journey. Tramping through the bustling urban landscape, we pass by the sometimes endearing but often tacky Queen Victoria Market, as the sun moves lower in the hazy summer sky without losing much of its intensity. The busy streets move slowly with creeping cars, and the passing trams are full to overflowing with those jaded commuters with tired eyes. It is better to be walking, we say to each other with a glance, even as we recognise the hint of petroleum in the air that fills our lungs. One feels the warmth of the pavement through one's shoes, the urban heat island effect experienced most directly.

Courtesy of Robbi Wymer © (http://www.robbiwymer.com.au/).

A thin veil of smoke also filters our view of the city, a result of unprecedented bushfires hundreds of kilometres away that began in late 2019 and continued through summer. These fires reduced to ash an estimated

11 million hectares (or 27 million acres) of Australian bush. Yesterday's January air quality in our most liveable city was officially 'hazardous' and the day before was ranked 'worst air in the world' by public health officials. Today's air is merely 'poor'; tomorrow's might be toxic. Welcome to the Anthropocene. We should be wearing breathing masks, like many of the people we pass. That said, we feel relatively responsible as we watch a man stub out a cigarette, throw it on the pavement, and immediately light a new one, relentlessly sucking tar into his lungs, personifying carbon capitalism. (Little did we know that in a couple of months we'd be wearing facemasks to protect ourselves from a different enemy.)

Countless lives have been devastated by these ongoing fires, human and non-human. The numbers almost defy human comprehension. It has been estimated that more than one billion animals have perished tragically, each of them innocent.[1] Some species are now freshly extinct. And it is *early* summer as we write these words, with the prospect or promise of fiercer furnaces lying ahead. The smoke in the city air reminds us of the intimate connection—barely a distinction—between town and burning country. The smoke has already reached New Zealand and NASA predicts the clouds of ash will do a full circle of Earth and return to Australia from the West.

Urban theorist Henri Lefebvre argued that social reality can no longer be neatly understood with categories like 'city' and 'countryside', but must now be analysed in terms of the complete urbanisation of society.[2] After all, the city is, and has always been, parasitic on that which is 'not city'. The smoke haze today affirms this connection and blurs seemingly commonsensical distinctions.

It is too late for ecological truths to be presented with kid gloves. It is time for radical honesty. An intensifying climate is turning this nation into a tinderbox that threatens to combust violently at the slightest provocation or accident, a comment that deserves both biophysical and cultural interpretation. What will next summer bring? As tragedies compound, perhaps it is time to say that the sun is setting not in, but on, the West. We are in the midst of a Great Endarkenment, our eyes still adjusting, as Australian Prime Minister Scott Morrison lauds his non-existent climate credentials and Donald Trump tweets threats of war to Iran. Even so, philosopher Georg Hegel once declared that the owl of Minerva only spreads its wings as the shades of night are gathering, implying that true knowledge only arrives at the end of a civilisational cycle.³ May we be so lucky. May insight arrive before it is too late.

Courtesy of Helen Lamb ©

We arrive at Flinders Street Station as the sizeable clock on its front face prepares to strike seven. We look up to see a grand building imposing itself on the landscape, not without a certain majesty. Do we look upward enough in the city? Are our urban goals sufficiently lofty? Dare we reach for the stars that are hidden by the light pollution of this great city? We urge you, dear reader, next time you wander the streets of your neighbourhood or city centre, don't forget to cast your eyes higher. Doing so can change the state of one's thoughts, drawing them higher too, as if by way of a current. Almost inevitably, however, the human situation and condition draw one's eyes and thoughts downward, the curse of fallen creatures. Our task is to resist and to demand more of ourselves.

The steps of Flinders Street Station are amongst the most fascinating places to watch the city of Melbourne unfold in all its spectacular diversity. The tramp and the *flâneur* do more than 'people-watch', but if you want to engage in the pastime of quiet observation in Melbourne, the steps of this central station (or any central station) are a prime place to do it. Cities are full of symbols and signs, waiting to be decoded, interpreted, and woven into new patterns of meaning and significance. For a time, we sit quietly and try to read our complex city.

According to Baudelaire, '[f]or the perfect *flâneur*, for the passionate spectator, it is an immense joy to set up house in the heart of the multitude, amid the ebb and flow of movement, in the midst of the fugitive and the infinite. To be away from home and yet to feel oneself everywhere at home; to see the world, to be at the centre of the world, and yet to remain hidden from the world'.[4] Your tramps are not mere idling *flâneurs*, but outside the station this evening our individuality fades to vanishing point as we find ourselves in a melting pot of business people and parents, lawyers and goths, preachers, backpackers, buskers, and police, and everything between and beyond. We take a moment to pat a dog that has kind, grateful eyes, while a sparrow nervously swoops nearby to collect crumbs from the city floor.

The unbounded variety of appearances amongst urbanites is astounding and somehow uplifting. Most fascinating of all is that the human experiment in living has arguably just begun. The deep future awaits, starting today—what will we make of it? As Thoreau once opined, 'there are as many ways [to live] as there can be drawn radii from one centre… [and human] capacities have never been measured… so little has been tried'.[5] He was cautioning us not to live our lives in the ruts of unthinking tradition and conformity, and instead to become something new, not merely a tired restatement history. Writing in the mid-nineteenth century, Thoreau was living in a time (like ours) of great economic transformation, and for him the railroad was the emblem of industrialisation. He often spoke of it metaphorically, as a representation of the emerging economic system that was fast changing the face of America and indeed the world. 'We do not ride upon the railroad', he concluded, 'it rides upon us'.[6]

Nevertheless, it appeared to Thoreau as if his neighbours had fallen into the common mode of living not because they preferred it to any other but because they honestly thought there was no choice left. We look out at our fellow city-dwellers on this hot evening and ask the same question. 'So thoroughly and sincerely are we compelled to live, reverencing our life, and denying the possibility of change. This is the only way, we say'.[7] Thoreau believed, however, there was more to life than the industrial city—and so do we. Or rather, we are exploring the possibility that there is more to life *in* the industrial city, a thesis we are testing in the hope of expanding the conditions for collective transcendence, as we sojourn through civilised life.

But the great clock at Flinders Station. We are drawn to the clock. A hurricane of busy people encircles us and then moves on with the traffic lights, a blur of noisy intention, but the clock, which we have each passed a thousand times, this evening invites us to pause, to step out of the rush, and to reflect on our temporality. What does it mean to live in this city of Melbourne during this third decade of the twenty-first century? The question has no single answer, obviously. Theoretical physicists report that something happened thirteen billion years ago, a singular event which a child with a limited vocabulary named the 'big bang'. Earth was formed around four and a half billion years ago, and perhaps a billion years later, the earliest signs of life sprung forth. Our species emerged around 200,000 years ago, and about 10,000 years ago, the Neolithic revolution marked the dawn of agricultural society. In 1859, the first oil well started being pumped in Pennsylvania, producing about twenty-five barrels each day.

And here we are now—in a world (before-COVID) demanding *100 million barrels of oil every day*—looking up at a clock outside a train station at the centre of this industrial city, listening to the complex beat of urban time passing in a globalised world. How is it that we live in a world with more advanced technology than ever before, and yet, as we read the latest IPCC reports, it is clear there is less time than ever?

No wonder the people leaving or entering the station look to be in an agitated rush. Everyone, it seems, needs to be somewhere other than where they are. Modern life defined.

Courtesy of Adrian Kenyon © (http://www.adriankenyon.com/).

How long have *homo sapiens* walked the land beneath our feet? Even today it is hard to comprehend the depth of Australia's human history. Billy Griffiths, in his book *Deep Time Dreaming*,[8] tells of how, when pioneering Australian archaeologists began serious work in the 1950s, it was generally thought that Indigenous Australian had only arrived a few thousand years before British colonisation. Carbon dating techniques have since pushed Australian history back into the mind-boggling expanse of deep time, with human arrival in Australia now estimated to be 60,000 years ago or more. It seems the so-called new world is in fact very, very old. That we so easily forget this reminds us that 'time has been a most effective colonising tool'.[9]

To refer to this period of time before colonisation as 'prehistory', as was commonplace not so long ago, is of course both absurd and arrogant. Archaeologist Vere Gordon Childe once remarked, 'it is not a sort of prelude to history but an integral part of history itself'.[10] Needless to say, history does not begin when white people arrive. And yet, here we are, your tramps, 'born of the conquerors' as poet Judith Wright puts it, in a 'haunted country',[11] feeling a deep connection with this great land but still learning how to belong.

We, the children of industrial civilisation, do not know how young we are. Our lack of historical appreciation is somewhat ironic given that the clock, as E. P. Thompson argued, was instrumental in organising and disciplining workers as industrial capitalism was born. Clocks would mark workers' hours of arrival and measure their idleness or efficiency in production. 'Time is now currency', Thompson said, 'it is not passed but spent'.[12] Of course there were sundials, water clocks, church bells, and crowing chanticleers to help regulate the day before industrial civilisation. But the mechanical clock, as Lewis Mumford contended, introduced a technique for time-discipline that made possible the idea of regular production, regular working hours, and a standardised product. Equally time can be enslaved to the cause of consumption. We think of a clock nearby at Southern Cross Station, where Old Father Time is suspended and upended by an illuminated red advertisement for money.

Paradoxically, then, the mass production of mechanical watches and clocks was both dependent on, and a product of, the mass production of watches and clocks. Indeed, the connection between clocks and capitalism was so well understood in previous eras that revolutionary activity has at times targeted clocks as being fundamental to capitalism.[13] Destroy the technology of time-discipline, and the capitalist cannot control the worker. The logic is too simplistic but it is understandable. Destroy all clocks![14]

Samuel Alexander © (http://samuelalexander.info/).

The shriek of a passing ambulance shakes us from our reverie and we find ourselves again sitting on the steps of Flinders Street Station in silence, beneath that instructive clock which surreptitiously disciplines all who accept its order. It is time to move on. Next to us a brood of youthful goths pass the time in peaceful coexistence in this most public of urban places. Why have they chosen these steps to gather? To be witnessed in order to feel alive? Perhaps they wonder of our purposes too—as do we. Uncomfortable glances are exchanged. We are all seeking, but seeking what?

Enchantment in the industrial city.

Notes

1. Sigal Samuel, 2020. 'A Staggering 1 Billion Animals Are Now Estimated Dead in Australia's Fires'. *Vox* (7 January 2020).
2. Henri Lefebvre, 2003. *The Urban Revolution.* Minneapolis: University of Minnesota.
3. Georg Hegel, 2005 [1896]. *Philosophy of Right.* New York: Dover, p xxi.
4. Charles Baudelaire, 1995. *The Painter of Modern Life and Other Essays.* London: Phaidon Press, p 9.
5. Henry Thoreau, *Walden*, in Carl Bode (ed.), 1982. *The Portable Thoreau*, p 266.
6. Ibid., p 345.
7. Ibid., p 266.
8. Billy Griffiths, 2018. *Deep Time Dreaming: Uncovering Ancient Australia.* Melbourne: Black Inc.
9. Barbara Adam, 2004. *Time.* Cambridge: Polity Press, pp 136–7.
10. Vere Gordon Childe, 1990. 'Australian Broadcasting Commission: Guest of Honour'. *Australian Archaeology* 30, pp 26–28.
11. Griffiths, *Deep Time Dreaming*, p 36.
12. Edward Palmer Thompson, 1967. 'Time, Work-Discipline, and Industrial Capitalism'. *Past and Present* 38, p 61.
13. See Jonathan Martineau, 2017. 'Making Sense of the History of Clock-Time, Reflections on Glennie and Thrift's *Making the Day*'. *Time and Society* 26(3), p 313.
14. Again, little did we know that the timelessness of the Coronaverse would soon be upon us, where the days of the week blurred into each other, disrupting timescales and temporalities.

5

Descent Pathways in a City of Gold

The machinic clock at Flinders Street Station had prompted reflection on our urban temporality; on the youthfulness of industrial civilisation juxtaposed against the depth of Australia's humanity. But still we seek a greater perspective on the city. What better place to get that perspective than from atop Melbourne's tallest building, the Eureka Tower, which boasts of having the highest viewing platform in the Southern Hemisphere.

The urban tramp doesn't just visit the 'grand landmarks' of the city, of course, and in fact will tend to deliberately avoid them in search of more singular and underground experiences, away from the tourists. By the same token, the tramp or *flâneur* resists convention and seeks to transgress boundaries, so our journey to the Eureka Tower is enticing just because it is something that the 'critical observer' of the city arguably ought to avoid. Just as the philosopher Michel Foucault once commented that to be anti-Nietzschean is, paradoxically, to be Nietzschean,[1] so too can we argue that it is in the spirit of the *flâneur* or the tramp to do things which they are not supposed to do. Thus, we make no apologies for our seemingly adolescent (but actually critically informed) defiance.

Courtesy of Mayara Menezes © (https://www.mayaramenezes.com/), Débora Matsuda © (http://www.behance.net/debmats), and Helen Araujo © (https://www.behance.net/profile/helenaraujo).

The evening is young and our sedentary day in the university office means that a further tramp through the city will certainly offer rewards for mind and body. Blood begins to flow and the middle-aged bones loosen as we walk south from Flinders Street Station. Before long we are descending a spiral staircase that delivers us to the southern bank of the Yarra River, known to the Wurundjeri people as Birrarung, 'a river of mists and shadows'. The river is fed partly from William's Creek, which now runs through a stormwater drain under the heavy bitumen of Elizabeth Street. At some point the creek must have been deemed inconvenient, so the early settlers developed it out of existence. A classic colonial move.

5 Descent Pathways in a City of Gold

What else flows hidden beneath this complex, troubled, but intriguing city? What is the state of Melbourne's urban subconscious? These, in part, are our research questions. In *Civilization and Its Discontents*,[2] Sigmund Freud argued that there are fundamental tensions between every individual's yearning for freedom and a society's expectations for conformity and the need to repress or sublimate anti-social desires. The result, Freud argued, is always unhappiness and neurosis. Does that explain the troubled mood we see subtly reflected in the eyes of our fellow city-dwellers (and each other)?

These uneasy thoughts arrive just as we watch people stream into the Crown Casino, an institution where the adrenaline of probably losing lots of money is mixed with the very tenuous prospect of winning lots of money. Maybe when these gamblers are rich they will be happy—but we have our doubts. Perhaps, at best, their neuroses will resolve into ordinary unhappiness. The words of neo-Freudian Erich Fromm remain as unsettling as ever:

> The fact that millions of people share the same vices does not make these vices virtues, the fact that they share so many errors does not make the errors to be truths, and the fact that millions of people share the same forms of mental pathology does not make these people sane.[3]

Most patrons will lose in the casino, of course, by economic design, resembling the capitalist mode of distribution. Even those who win the rat race too often live like rats to get ahead. Surely, we can do better. What limited imaginations can only find meaning and entertainment in gambling and accumulation? Why must we always struggle for more, and not sometimes be satisfied with less? Our challenge, it seems, is to resist the 'pathology of normalcy'.[4] We bypass the casino without regret, beckoning industrial civilisation to follow.

Southgate, where we walk, is one of the most densely populated areas of Melbourne and one dominated by high-rise development and riverside restaurants. Our city, like all cities, is full of contractions. This evening the almost mystical light of dusk filters our experience as we step closer to the water's edge. It is good to wander the city, to partake in its ebb and flow. There is resounding beauty amongst all the ugliness and ordinariness, if only we slow down enough to see it. For a moment we let ourselves appreciate the spectacle of this mighty river, older than humanity, as the night lights of the city begin to reflect off the shimmering water. Behind us the hum of the city vibrates with life. Again, we look up to orientate ourselves. Our destination is towering nearby.

We had planned to make this vertical journey up Eureka Tower, so earlier in the day we purchased tickets, like good tourists and consumers. We sense Baudelaire turning in his grave, but nevertheless we let his disappointment wash over us. A tidy $23 dollars each to get up to the Skydeck and a further $12 each to experience 'the Edge'—a room that extends from the 88th floor of the Eureka Tower and suspends you 300 metres over Melbourne in a glass cube, including a clear glass floor.

Courtesy of Helen Lamb ©

Surely these ticket prices would be prohibitive to people and families living from pay cheque to pay cheque. It is a specific example of capital's tendency to insidiously segregate people and rule over urban space based on 'ability to pay'. What of the right to the city? What of the right to experience its delights even, or especially, if one is poor? Experiencing the Skydeck could be shared freely—at 'zero marginal cost'—without diminishing the gift. And yet, security guards are present, hidden in plain sight, to ensure only those with enough money are permitted to enter. Their smiles disguise the fact that if you do not pay, they will remove you, forcibly if necessary, a quiet reminder of the intrinsic connections between private property, power, and the thinly veiled threat of violence. These reflections simmer uncomfortably as we enjoy the privilege of being elevated 88 floors in the air.

The lift opens and we navigate ourselves to the main viewing deck. Smoke haze aside, it is a clear night, and we are struck by the breathtaking vista of our luminous and radiant city. We are reminded of 'The City of Gold', a photograph by Andy Serrano that captures the vast, glowing, infinitely sprawling landscape of a metropolis at night. Before us a constellation of lights from indistinct cars and buildings shift slowly, a cluster of bright stars contrasted against the dark ether of night, out to the horizon and beyond. Where does this urban footprint end? Does it end? What would a fully developed world look like?

Courtesy of Andy Serrano © (http://andyserrano.deviantart.com/).

The panorama is quite majestic from this distance at night, enchanting even. 'To be enchanted', Bennett says, 'is to be struck and shaken by the extraordinary that lives amid the familiar and the everyday'.[5] The view from up here is, if nothing else, extraordinary—not quite human in scale, almost divine. At the same time, it is deeply disturbing to those who know what it means, or might mean, in terms of how growing portions of humanity have come to inhabit our fragile planet in this increasingly urban age. All those lights, beautiful though they are, signify the energy intensity of our infernal civilisation like nothing else.

The problem with this stunning but unsettling view of the city is that it highlights our dependency on fossil energy, which built the skyscraper in which we stand and which has given rise to an energy crisis with two worrying dimensions. First, fossil fuels are finite resources that are being consumed at dizzying rates, such that their supply will one day peak and decline even as demand promises to grow. In early 2020, as we have already noted, the world was consuming 100 million barrels of oil every day, growing every year… while stocks last. The US shale oil boom, which has kept oil prices under control in recent times, is expected to plateau in the next few years (and the COVID-19 disruption, which arrives in a couple of months, could well be shale oil's early death knell).

So, as we look out over this petroleum-dependent city we ask: exactly where we are going to get the oil to meet industrial civilisation's ever-growing demand? There are more than one billion carbon-dependent vehicles on the road today, and the International Energy Agency estimates that in a couple of decades we'll be approaching two billion. A recent HSBC rewport projects that the world will need to find the equivalent of four new Saudi Arabias by 2040 to meet projected demand.[6] So far, we have only found one Saudi Arabia, even though the oil industry has been ferociously searching for more but finding less. While 'peak oil' concerns may have been casually dismissed by cornucopian analysts today, rumours of the death of peak oil have been greatly exaggerated. Peak oil is not dead, we surmise; it is only in remission. Watch this space, or better yet, plan for it. Our view from up here confirms the premise that we inhabit an industrial city, hungry for fossil fuels.

Peak oil, however, is no longer the primary or most pressing concern for carbon civilisation. The other element of the global energy crisis is that the combustion of fossil fuels is the leading driver of the climate emergency, meaning that humanity must decarbonise *by choice* even before we are forced to do so through geological depletion. Peak oil now! Further to that seemingly insurmountable challenge, it remains highly uncertain whether renewable energy technologies will be able to fully replace the nature and extent of energy services provided by fossil fuels in an energetically or financially affordable way. It is unsettling to read the *World Energy Outlook* 2018 and discover that wind, solar, and geothermal together only make up around 1.7% of global energy supply—making a 100% post-carbon transition seem truly daunting.[7] Try building the Eureka Tower, or flying the dozen passenger planes we see scattered across the heavens this evening, using only solar, wind, and biofuels.

If we are serious about managing geological depletion and keeping within humanity's fast-shrinking carbon budget,[8] then humanity—starting with the world's highest energy-using nations (like Australia)—needs to learn how to consume much less energy, not just assume we can 'green' existing let alone growing supply. All this makes it plausible that the urban industrial future will be defined by increased energy scarcity not energy abundance, which implies an 'energy descent' future with rising energy costs relative to today. Assuming the Eureka Tower is still standing in one hundred years, what will our view look like then?

Although the Yarra River will still flow, we suspect there will be fewer planes in the sky and fewer cars on the roads, replaced with more cyclists and walkers. There will be reduced light, noise, and air pollution, as wastefulness and combustion become increasingly unaffordable. More abstractly, given the close connection between energy and industrial production, the declining energy availability and affordability will likely lead to economic contraction and reduced material affluence.[9] But there is no reason, we feel, that this more humble energetics and economics of sufficiency would be inconsistent with a 'frugal hedonism'.[10] There is the potential of a prosperous descent ahead, if only *homo urbanis* is able to untangle the many contradictions we face today and show the courage to plan for economic contraction thoughtfully and with compassion, before it arrives with furious surprise.

Blinded by techno-optimism, most people do not entertain these possibilities, at their own peril. For politicians, this post-industrial vision is unspeakable. Michael Moore's 2019 documentary *Planet of the Humans* is crude and clumsy, but it makes an essential point few dare to make: technology cannot save capitalism from cannibalising itself. The growth economy has no future. No wonder so many people are upset by the film. It's blasphemous.

We escape our thoughts for a moment and privilege our eyes: few would deny the city from up here looks sublime. There is a buzz of excitement around us as people from all around the city and indeed the world drink in the atmosphere, 88 stories high, joining us in a state of poetic alienation. Still, if energy descent lies in the future, does this glowing urban landscape before us have a future? Might there need to be another Eureka Rebellion—a rebirth of democracy—if this city of gold is to be transcended without tragedy?

Perhaps this speaks to the third element of the energy crisis: the lacklustre energetics of ethics in the world today, or the absence or scarcity of what Bennett calls 'affective propulsions'.[11] We all know that the Age of Oil is a brief anomaly, a one-off inheritance of dense fossil energy that is being spent, as if there were no tomorrow, with reckless abandon. We all know that our destructive economies are destabilising climatic systems and causing a holocaust of wildlife and biodiversity. But accepting the 'moral code' of deep decarbonisation is necessary but insufficient. What is also needed is an embodied sensibility, which organises and informs our social and political tastes, feelings, and affects into a coherent ethical energetics that generate the impetus to *enact* decarbonisation. As Bennett argues, moral codes 'remain inert without a disposition hospitable to their injunctions, the perceptual refinement necessary to apply them to particular cases, and the affective energy needed to perform them'.[12]

Just as fossil fuels are finite, so too is industrial civilisation, including its urban forms. Perhaps we already live in the autumn of carbon capitalism. Everywhere dead leaves are falling as civilisation enters a long emergency, a sign that change is coming, whether by design or disaster. But what comes next seems unknown—as the coronavirus will soon teach us rather starkly. Perhaps, hopefully, the future is still unwritten, and therefore unknowable and subject to human influence. The only thing that is clear is that if humanity does not remove its rose-tinted glasses our urban futures will surely come to us by shock and surprise. The future is not what it used to be, nor are we, *homo urbanis*, who are writing it. All is flux. One never steps into the same city twice.

After a time, the enchantment of the view from the Skydeck fades; its impact is lost. This is the way with enchantment, she is a fickle lover who comes and goes as she pleases. A hint of boredom dawns. Psychologists might diagnose this phenomenon as 'hedonic adaptation', but fear not, we have tickets to 'the Edge' to reignite the consumer experience in search of lasting satisfaction. Soon enough we find ourselves at the doorway of the glass cube protruding from the top of this skyscraper. We prompt each other to enter first, to ensure it is sturdy. Pretending to be brave we step forth upon the transparent floor, 300 metres above the indistinguishable pedestrians going about their lives on the ground.

For the time being, at least, the floor holds, but it is clear there is not much stopping a great fall—it feels like something small, even a microbe, could take us down. We feel like Coyote in the old cartoon *Road Runner*, who would sometimes fail to stop himself before running over a cliff's edge. The joke was that he would not fall until he looked down and realised that there was nothing holding him up. The reference might sound trite or lightweight, but it has a deeper resonance. Are we, participants in industrial civilisation, already over the edge? Dare we look down through the thin layer of substance holding us up? Admittedly, it can be frightening. The prospect of a fall is no joke.

There are, to be sure, many descent pathways that lie ahead (or below) for this city of gold and its inhabitants.[13] Which will we choose? Or will a path choose us? This evening we are fortunate enough to be able to take the lift down and a certain security returns as we step out onto the solid pavement, reentering the urban fray on the ground level. The night is warm, so we decide to stroll home, over Princes Bridge, northward through the city centre up Elizabeth Street, stopping off at the Last Jar for a cold ale before parting ways to our respective abodes. We watch the trams move noisily past as we enact our freedom by choosing to walk, slowly enough to draw in a multisensory experience, an intentional method to 'get to know' the city in a different way, to decipher its text and take its rhythms less for granted. There is much in the city that exists and is real but cannot be seen.

Beneath us William's Creek still flows, but as a hidden drain. It speaks to a subterranean world beneath all industrial cities, of watercourses and their histories, contained and drained by engineering that sought to make humans safe. We continue to float eerily upon these buried natures. In Melbourne, we continue to deny their ancient ownership and curation by Indigenous peoples. One day, perhaps, it might all burst to the surface and surprise us.

Notes

1. Michel Foucault, 'Return of Morality' in L. Kritzman (ed.), 1990. *Michel Foucault: Politics, Philosophy, Culture*. New York: Routledge.
2. Sigmund Freud, 2010 [1930]. *Civilization and Its Discontents*. London: Penguin.
3. Erich Fromm, 2006. *The Sane Society*. London: Routledge, p 15.
4. Ibid., p 12.
5. Jane Bennett, 2001. *The Enchantment of Modern Life: Attachments, Crossings, and Ethics*. Princeton: Princeton University Press, p 4.
6. Kim Fustier, Gordon Gray, Christoffer Gundersen, and Thomas Hilboldt, 2016. 'Global Oil Supply: Will Mature Field Declines Drive the Next Supply Crunch?'. HSBC Global Research Report (September 2016).
7. See generally, Samuel Alexander and Joshua Floyd, 2018. *Carbon Civilisation and the Energy Descent Future: Life Beyond This Brief Anomaly*. Melbourne: Simplicity Institute.
8. For a recent paper outlining the magnitude of the climate challenge, see Kevin Anderson, John Broderick, and Isak Stoddard, 2020. 'A Factor of Two: How the Mitigation Plans of "Climate Progressive" Nations Fall Far Short of Paris-Compliant Pathways'. *Climate Policy*. https://doi.org/10.1080/14693062.2020.1728209.
9. See Alexander and Floyd, *Carbon Civilisation*.
10. Annie Raser-Rowland and Adam Grubb, 2017. *The Art of Frugal Hedonism*. Hepburn Springs: Melliodora.
11. Bennett, *Enchantment*, p 3.
12. Ibid., p 131.
13. David Holmgren, 2009. *Future Scenarios: How Communities Can Adapt to Peak Oil and Climate Change*. White River Junction, VT: Chelsea Green.

6

Adrift in the Devil's Playground

Your tramps are on a ramble in the central business district. We stop near RMIT University at its northern edge. On a backstreet, we encounter a playground heaving with young adult bodies. This is the RMIT A'Beckett Urban Square, with its basketball and futsal (mini soccer) courts, today arrayed with excited players. This precious and very human space is closely surrounded by new and constructing residential towers but boils back against all this vertical monstrosity with latitudinal life force. Right now, it is an urban field of happy players. Games in the city… other thoughts come briefly to mind, the 'development game', in particular, that has long been played in Melbourne.

Like all capitalist cities, Melbourne has always thrived on what Leonie Sandercock memorably described as 'the land racket'.[1] The racket is speculative urban development driven by thirst for the super profits that so often reward the canny (and occasionally the crooked) entrepreneur. Why a racket? Some prefer the softer 'development game'. Both tropes tell us that the urban process of change in market societies is a kind of competition far removed from the textbooks of Chicago school economics, where loose rules and blind-eyed umpires favour the boldest not the fairest players, whose many stripes include shrewds and swindlers.

Melbourne has its colourful land racket past, beginning with the swirling and ultimately self-destructing land booms of the 1880s, described elegantly in Graeme Davison's *The Rise and Fall of Marvellous Melbourne*.[2] Michael Cannon, another historian, described the entrepreneurs, great and small, behind the city's explosive development surge in those years as *The Land Boomers*.[3] Later we shall meet one of those Boomers, the appositely named (Sir) Thomas Bent (see our chapter 'The Monumental Army'), whose life effort was to curve the rules of the game, as well as the resources of the watching public, to his own pecuniary interests. As with all capitalist free-for-alls, this one ended in tears; a calamitous market crash from 1890 bringing many of the boomers back to earth (quite literally in the sad cases of those who responded to ruin by casting themselves off some of the city's taller buildings).

Brendan Gleeson ©

As they say about all falls, it's not the plummet that kills you but the sudden stop at the end. Many of the bright lights of Marvellous

Melbourne were extinguished by the bust. The city wore sackcloth and ashes for a long time after this cataclysm. A rather grey period of austerity and good behaviour attended in its wake for many decades. But by the 1970s the players had returned to the greenfields of new development opportunities, mostly at the metropolitan fringe. Planning laws, such as they were, suffered subversion, while public funds for housing development were effectively misappropriated by land swindlers and other fast players. This was more Hunger Games than a fair match; the vigorous futsal enthusiasts in front of us play much more fairly and pleasantly than did the crooked developers and officials in the 1970s.

Courtesy of Jorge Alaminos Fernández © (http://litoralgrafico.tumblr.com/).

Eventually the scams came to light and a 'land scandal' was declared by the press and some political interests. A small cast of minor villains were whacked with wet pillows in prolonged and halting investigations. It was widely held that the main boomers escaped the field, ignoring the umpire's faltering whistle and slipping away before the siren. Inquiries were held and the development community was put on a good behaviour bond. The land racket subdued and resumed its more typical modus of rank speculation and opportunism, staying below the line of hard corruption. Ah the rise and fall, the rise and fall... there is something hypnotic about the oscillating pattern of villainy and venality in urban capitalism.

Meanwhile one of the urban tramps has fallen into a golden slumber, on a warm step by the RMIT A'Beckett Urban Square, lulled by its happy disports. The other tramp wanders for a time, through alleyway and backstreets, with purposeless purpose, ruminating like a Thoreauvian camel, chewing slowly over some inner work and travelling well beyond the marketplace despite never leaving its depths. Both tramps are practising post-capitalism in the midst of the neoliberal order, an art that is never mastered. As for think-walking, Rebecca Solnit describes the transgressive act with characteristic insight:

> Thinking is generally thought of as doing nothing in a production-oriented culture, and doing nothing is hard to do. It's best done by disguising it as doing something, and the something closest to doing nothing is walking. Walking itself is the intentional act closest to the unwilled rhythms of the body, to breathing and the beating of the heart. It strikes a delicate balance between working and idling, being and doing. It is a bodily labor that produces nothing but thoughts, experiences, arrivals.[4]

The walking tramp returns to find the other still practising post-capitalism—asleep, daring to be idle in this industrious city. He wakes from his reverie in a sunny spot, the day's newspaper on his head, only to stare at new headlines reporting here in the now of early 2020, a new upwelling of land racketeering. Ah, but our shock is relieved by a comfortingly familiar tale of graft at Melbourne's creaking urban fringes, this time in the City of Casey, a municipality to the southeast of the city centre. The State's corruption watchdog (a rather friendly toothless little fellow), the Independent Broad-based Anti-corruption Commission (IBAC), is conducting hearings into what emerges as a full-bodied tale of development graft with rich, almost pleasing, notes of innovative villainy. Amongst the latter we count the setting up through generous commission, by boomers, of a bogus 'community' group that strangely had a supportive, indeed misty eyed, view of local developer ambition. We dip our lids, as our fathers would have said, to IBAC and its laborious investigations. May it eschew the wet pillow and bring justice to the field. We shall not take our hats off until this is the case.

This brings us to the question of the land racket, which we know as an endemic not pandemic feature of capitalism. Despite the historical testimony, and what many think, the game survives all scandals and censures and finds endless means for survival under ever-evolving institutional and social conditions. The permanent lure of speculative profit is just too great to be contained by easy regulation even with periodic tightening. Casey's corruption scandal reminds us that all is not smooth at the city's fringes. But the truth is that most contemporary suburban development is probably compliant, not corrupt, which means it satisfies the weak regulatory order maintained by thirty-one metropolitan councils and a lethargic state government. Land grabbing is an ordinary thing in the capitalist city.

So, where is the beast at work, as we tramp and ponder today? Enter for our thoughts, the compact city, an ideal that has dominated the imaginaries of planners (and increasingly developers) for decades in western cities. Its grip in Melbourne is evoked by a series of well-meaning decadal planning strategies that have sought to reinstate the honourable town planning idea of containment. Even given weak regulatory will to enforce the ideal, it has progressively captured the imagination and ambition of the whole development game team (planners, developers, commentators, advocates). The consequent slow arrest of development on the fringe (though still subject to corrupting greed), combined with a willingness of the state government to relax (and effectively deregulate) planning controls in the inner city, has produced a new development surge that departs from the old. The new easy is the inner city.

As we tramp like lost ants in the shadow of countless protruding towers, we get a sense this urban landscape has lost its human scale. We note with Lauren Elkin that 'what we build not only reflects but determines who we are and who we'll be'.[5] This insight is a two-edged sword, raising concerns about what we've already built and how it is shaping us, but also providing hope about who we could yet be, if we built otherwise, privileging people and planet over profit.

For the past decade or more, under conditions of convulsive population growth, Melbourne's inner city and increasingly its surrounds have become a new playing field for the development game. The game has moved from greenfields to brownfields. From the green turf of football fields to the hard tack of futsal courts. An avalanche of development capital has poured into the 38 sq kms that constitutes the core municipality, the City of Melbourne. Earlier we mentioned a metropolitan growth figure of 500,000 persons for the last five years. Much of this has been inwardly focused. There are now over 70,000 overseas higher-education students living or studying in the City of Melbourne alone (numbers that will have to be readjusted, of course, in and for the after-COVID era). To tramp the contemporary CBD is to wend a way through the rising canyon depths created in recent years by freewheeling tower developments. Our present perch by the playing ground is at the base of a new gorge of rising urban ambition and value.

Our memories hark back to a 2015 study of inner-city development by Leanne Hodyl, a Churchill Fellow, which was briefly but widely reported in the media.[6] This powerful work disavowed by careful comparative analysis the laissez faire planning culture presided over in the inner city by a succession of state governments, most especially the incumbent Planning Minister at the time of its composure, Matthew Guy (2010–2014). Through the years of his incumbency Guy approved many arguably inferior gargantuan tower developments for the city, attracting much criticism from various quarters along the way. But Guy remained impressively insouciant, ticking off a vast tide of vertical development that is still remaking the city today and will continue to do so for years.

6 Adrift in the Devil's Playground 73

In 2020, the development approval and construction visualisation provided by the City of Melbourne's 'Development Activity Model' shows the full scale and legacy of Guy and other system authorisers.[7] Using this web tool to visualise where we now stand shows a nightmarish gorge of closely packed and poorly conceived towers, built and awaiting construction. As Hodyl pointed out, this scene is quite unlike anything else in the world. We might describe it as the 'Elizabeth Street Valley of Death (to urban amenity)'.

Brendan Gleeson ©

Leanne Hodyl's study compared Melbourne's cowboy planning framework to the more settled, publicly interested systems in Hong Kong,

New York, Seoul, and Vancouver. Hodyl demonstrated through a small set of illustrative case studies just how extraordinary and greedy were the development densities allowed in Melbourne compared to the other cities of her study. She showed by the numbers just how extraordinary were Melbourne's 'hyper-dense, high-rise' developments and block aggregations. Melbourne has been allowed to suffer extraordinary high-rise densities which would never be allowed in other global cities otherwise renowned for their densities.

Stirring ourselves at the A'Beckett Urban Square perch, your tramps realise we are on the edge of one of Hodyl's two Melbourne case study sites. The Square sits in the corner of a city block that Hodyl describes as having one of the highest allowable built densities in the world. We look up into a cluster of dark and mirrored beanstalks that seem to reach into the clouds. An unpleasant augury attends this urban unsettlement. Might a huge angry giant climb down these glittering stalks to smite the Lilliputian hordes below? Our happy futsal congregation? Must they pay for all this towering urban wreckery?

The tramps are tramps and need no urban high life. But these towers take the cake for their soaring decrepitude. Speculative development is often obvious for its capitalist featurism—for design and construction qualities especially that imbibe the spirit of windfall profit not public or even resident interest. We can think of 'Day 1' shoddy features; poor-quality external makeovers and internal fit-outs that breezily passed the test of a privatised 'building inspection system'. It all seems on display or promised here.

We leave A'Beckett Urban Square with a sense of new injury. A'Beckett Urban Square is a 'sacrificial' development site; its happy courts will at some point, dictated by the market, give way to an approved towering monster. The development game trumps all other games. As the Casey story reminds us, the old horizontal development racket with all its depredations continues quietly, and sometimes not, in the outer fringes of the city. But this here in recent years is something new.

How to describe the new vertical misanthropy that literally overshadows Melbourne, and many other cities besides? The term hypertrophy comes to mind. We borrow the word from medical science to describe a sick urban state. This new term, urban hypertrophy, denotes that city regions are rapidly growing under the impress of high population growth but are also experiencing internal 'cellular' enlargement (hypertrophy) as built structures and centres increase in scale and density. It is a cancer in the heart of our city, spreading now outwards as development opportunity and ambition combine in wider spatial arcs.

In the (erstwhile) world's most liveable city we find ourselves, during all this chorused virtue, adrift in the Devil's Playground. We prefer the (not so) quiet virtue of futsal, an enchantment strategy that at least gives expression to the sense of play. Disenchantment certainly comes easy in the industrial city. Bennett's thesis regarding modernity is being played out here and now in this towering urban landscape, given that 'the very characterisation of the world as disenchanted ignores and then discourages affective attachment to the world'.[8] That makes the telling of alter-tales all the more important, since 'the mood of enchantment may be valuable to the ethical life',[9] evoking a presumptive generosity of spirit, and one strategy to 'enhance the enchantment effect is to resist the story of disenchantment of modernity'.[10] At least, let us all be open to moments of enchantment in the industrial city—like watching an amateur game of futsal in the sun—and allow the marvellous to erupt in the everyday, even in a city, so-called Marvellous Melbourne, that too often marches to the heavy drumbeat of disenchantment.

Notes

1. Leonie Sandercock, 1979. *The Land Racket: The Real Costs of Property Speculation*. Kuala Lumpur: Silverfish Books.
2. Graham Davison, 1995 (revised edn). *The Rise and Fall of Marvellous Melbourne*. Melbourne: Melbourne University Press.
3. Michael Cannon, 1973 (reprint edn). *The Land Boomers*. Melbourne: Melbourne University Press.
4. Rebecca Solnit, 2000. *Wanderlust: A History of Walking*. London: Penguin, p 5.
5. Lauren Elkin, 2017. *Flâneuse: Women Walk the City in Paris, New York, Tokyo, Venice, and London*. London: Vintage, p 33.
6. Leanne Hodyl, 2015. *To Investigate Planning Policies That Deliver Positive Social Outcomes in Hyper-Dense, High-Rise Residential Environments*.

Report to The Winston Churchill Memorial Trust of Australia. Unpublished mimeograph. Melbourne.
7. City of Melbourne, 'Development Activity Model'. https://www.developmentactivity.melbourne.vic.gov.au/ (accessed 20 June 2020).
8. Jane Bennett, 2001. *The Enchantment of Modern Life: Attachments, Crossings, and Ethics*. Princeton: Princeton University Press, p 3.
9. Ibid., p 4.
10. Ibid.

7

Grave Matters: Death in the Liveable City (Part I)

If there's one thing you don't hear much about in the constant chorusing of praise for the 'World's (no longer) Most Liveable City', it is death. Echoing this, Melbourne's sprawling cemeteries don't seem to get much airtime in a city otherwise obsessed with its booming real estates. But we know that in capitalism boom is inevitably punctuated by bust, just as life is by death. This silence on the cycle of things beckons our attention. To resist the death drive, we decide on a cemeteries tramp. This must follow a geography of the city's historical development, visiting the encampments of the dead that shadowed the armies of the living in the long outward trek from its original centre to its ever-spreading suburbia. Over the centuries the dead have sprawled as much as the living. Spectres at the feast of urban growth.

We begin at the city's first formal burial ground established in the immediate wake of the European invasion, recalled as The Old Melbourne Cemetery. It was opened for business in 1837 on the site of what is now the Queen Victoria Market, which was established in the late 1860s. Both coexisted for a time; the receipt of earthly remains accompanying the selling of earthly produce until the cemetery was finally absorbed into the market in 1917.

Courtesy of Alessandro Gatto © (http://www.alessandrogatto.com/).

We arrive at the eastern edge of the old cemetery, long carpeted by a large asphalted carpark. Here is a monument erected in 1881 by an adoring citizenry to honour the city's 'founding father', John Batman. After his death from syphilis in 1839, the 'father' was himself installed in the cemetery, later (in 1920) removed with other 'pioneers' to one of its suburban successors (we'll return to this story later). Despite his ennoblement in Victorian public memory, Batman, a Tasmanian grazier, was less noble adventurer and more ambitious entrepreneur, looking to the Port Phillip settlement as means for expansion of his business. He negotiated the original and only treaty with the local Aboriginal peoples in 1835, a transaction that was not to their advantage. His avarice had form. Wikipedia reminds us that the artist John Glover, Batman's neighbour in Van Diemen's Land, 'said Batman was "a rogue, thief, cheat and liar, a murderer of blacks and the vilest man I have ever known"'.[1]

7 Grave Matters: Death in the Liveable City (Part I) 79

With this in mind, we blanch at a rogue's memorial which avers he had founded a settlement 'on the site of Melbourne *then unoccupied*' (our emphasis). Life eventually caught up with Batman, his death followed by interment in this place, but it was a long time before the life of Melbourne's *terra nullius* settlement myth came to an end. We note a small plaque that was installed on the base of the monument in 2004 by the city of Melbourne which coyly describes the settlement lie as 'inaccurate' and attempts an apology to the traditional owners. The 'historically corrected' memorial still trumpets the legacy of the 'founding father' to passers-by, mostly shoppers, loaded with produce, heading for their cars. As Australian artist Ben Quilty observed recently, 'We're good at building memorials for people, if you're white'.[2]

Next stop on our post-liveable city tour is the Melbourne General Cemetery, draped like a beautiful deathly crown on the northern perimeter of our employer, the University of Melbourne (1853). Unlike the more rudimentary and crowded old cemetery, this place of rest was intended when established in 1853 to embody (so to speak) the new Victorian principles of sanitary and respectful interment. Just as the suburban blocks of the living were to grow in size, so too did the plots of the dead, in new cemeteries such as this which observed the spacious principles of the botanical garden (see also, St Kilda to the south [1855], Williamstown to the west [1858], Kew to the east [1859]). The ever-simmering sectarianism of the living was also mirrored in a strict spatial segregation of the dead along denominational lines.

One hundred and sixty-six years of growth has rendered this boneyard a fine place indeed, its many thrusting monuments to wilted grandees alongside the humble flattened graves of common folk. Mausoleums, chapels, gardens, sweeping drives and special memorials constitute a fine estate. A folk rococo Elvis monument brings us up towards the present.

This cemetery has life in it yet and continues to receive the departed. This is boldly signalled at its main southern entrance where we encounter on a windy Summer day a long flapping banner announcing the happy news of a 'Limited Release of Exclusive Graves. Enquire today…'. Further along the busy road that skirts the cemetery another banner gushes, 'Final Mausoleum. Now Open'. The passerby is encouraged to 'Secure your position today'.

Brendan Gleeson ©

We are impressed, if a little unsettled, by this realty of death. It is strangely invigorating—almost like the energy ones gets from reading the seemingly morbid existentialists like Jean-Paul Sartre or Albert Camus, who tirelessly focus our attention on death, but with purpose, not self-indulgence, in order to remind us all of life's finitude, of its humble majesty, and the heightened responsibility of grasping life and squeezing more out of it than we could have ever imagined. Do not worry so much about seeking the status of an 'exclusive grave'. Just exhaust one's earthly project with passion and compassion, and any afterlife will take care of itself.

It is curious how a focus on death can breathe life into the human condition. It was this insight, of course, that led philosophers in earlier ages to write with a skull on their desk, looking them in the face, a constant and provocative reminder that life is short and too precious to

7 Grave Matters: Death in the Liveable City (Part I)

waste. This vital lesson is easy to forget—but not in a cemetery. Today the reality and realty of death evokes a mood that experiential philosopher Philip Fisher would describe as 'pure presence'.[3] It would not surprise Bennett to hear that we found ourselves noticing new colours, discerning details previously ignored, hearing extraordinary sounds, as familiar landscapes of sense were sharpened and intensified by coming face to face with our mortality in this disconcerting way. Bennett says that enchantment includes this 'condition of exhilaration or acute sensory activity',[4] and that 'to be simultaneously transfixed in wonder and transported by sense, to be both caught up and carried away – enchantment is marked by this odd combination of somatic effects'.[5]

In any case, the moment breaks the spell of speeded timelessness that permeates the everyday life of the industrial city, reflected in the seemingly endless manic traffic that roars and honks behind our backs. The good retail work of the cemetery management reminds us that time and space eventually run out, even for the dead. Secure your plot now. This feels too sombre, so we allow a smile observing the flapping banners which evoke, perhaps more than anything else, Melbourne's long and shameless obsession with real estate. A long history of booms and busts, peaks and troughs, buildings and graves. A kind of 'upstairs, downstairs' urban tragicomedy.

As we noted, Batman's adventuring did not conclude with his death. Just as his short eventful life (thirty-eight years) was disrupted by an early demise, his long (aren't they all?) death was disturbed by the removal of his remains, with 913 other old colonists, to one of the newer cemeteries whose establishment marked the trek of suburban expansion in the twentieth century. Fawkner Memorial Park was opened in 1906 and sprawls across an impressive 111 hectares. It is indeed a suburb in its own right, with a train station by its side which once received funerary trains, a particular transport innovation and convenience of Victorian urbanism that, like the dead, is now sadly departed. Consultation of the Park's website reassures us, however, that the spirit of capitalist change (for change's sake) runs like a red thread through the business of death: 'Fawkner Memorial Park has a history of innovation and leadership in the Victorian cemetery industry'. We are enchanted again, struck by an unsettling mood which we do not fully understand.

Being interested in urban disturbance we journey to pay our respects to the dislocated old colonists, resting since 1920 in a spacious rose-bedecked reserve within the Fawkner estate. First to our attention, another monument to that foundered father Batman! As Yeats enjoined for his own grave, *Cast a cold eye...horseman pass by!* We do. And then what a splendid place this proves to be. The records attest that in their original resting place, the old colonists (mostly buried in the 1840s) were crowded together like early Victorian slum dwellers. They were richly repaid for their disturbance in 1920 by resettlement in spacious and neatly aligned grave rows, original monuments respectfully reinstalled. All this rousing of the dead: *You cannot rest, we are not finished with you!*

These old colonists were the original Progress Association of Victorian colonial capitalism. Headstones like fetishes to the cause of industrial growth. Inscriptions that recall the countless sad misadventures of the footsoldiers (and their families) of early mercantilism; the children taken in swathes by insanitary conditions, the young adults extinguished by accident or the early wearing out of life in a stolen, not yet understood land; the 'aged pioneers' finally succumbing after outliving children and relatives. It reminds us how death so often surprises and mocks the steady if brittle calculus of the industrial city. Here on headstones are the stories of deathly enchantment. We enlist the long dead Sydney poet Kenneth Slessor who evoked,

> ...tablets cut with dreams of piety
> Rest on the bosoms of a thousand men
> Staked bone by bone, in quiet astonishment
> At cargoes they had never thought to bear,
> These funeral-cakes of sweet and sculptured stone.[6]

Again, and to close, let's relax the grip of certitude, of sombre realisation. Amidst the text of remembrance etched on headstones we find, unexpectedly, ambivalence. But of course, why are we surprised? Freud diagnosed this as a deep whisper in the soul of modern sensibility. We hear its early utterances in the headstone inscriptions. Thomas Armistead who 'DEPARTED THIS LIFE Dec—15—1850, AGED 34 YEARS' was (in old English text) 'Deeply regretted'. Poor James Purse was drowned crossing the Merri Creek (runs north of Melbourne) on

'28th day of Nov'r 1849, AGED 28 YEARS'. Alive he was 'HIGHLY RESPECTED BY ALL WHO KNEW HIM'. At death he was 'DEEPLY REGRETTED BY ALL WHO KNEW HIM'. Do we sense, even from this historical distance, that imperfect battle always raging between love and hate, still leaving its monumental traces before us today? No bodies, including your tramps, stand apart from this question.

We mean no disrespect with our enchanted journeys through the suburbs of death. Quite the opposite; our intention is to bring to life, and thereby honour, the morbidity of the growth machine of capitalism. The dead march with us as we journey to the netherworlds of material (im)possibility.

To finish the journey, last words to Charles Baudelaire, fellow observer of city life. Like the 'innovators' of Fawkner Memorial Park, he saw death as journey not closure. Not the accountant's final depreciation, rather the natural revaluation of all that has been extinguished by industrial imposition. Industrialism desires deathless repetition: we are borne to imagine there is nothing new under its hostile sun. Until, inevitably, we arrive, so often surprised, at the moment of earthly embarkation.

> …It's time. Old Captain, lift anchor, sink!
> The land rots; we shall sail into the night;
> if now the sky and sea are black as ink
> our hearts, as you must know, are filled with light.
>
> Only when we drink poison are we well—
> we want, this fire so burns our brain tissue,
> to drown in the abyss—heaven or hell,
> who cares? Through the unknown, we'll find the *new*.[7]
>
> – Baudelaire, *The Voyage*

Notes

1. See https://en.wikipedia.org/wiki/John_Batman (accessed 9 June 2020).
2. Ben Quilty, 2020. 'Ben Quilty's 2020 Vision: Commemorate Massacres for Sanity, Goodness, and Healing'. *The Guardian*, 5 February 2020.

3. Philip Fisher, 1998. *Wonder, the Rainbow, and the Aesthetics of Rare Experience*. Boston: Harvard University Press, p 131.
4. Jane Bennett, 2001. *The Enchantment of Modern Life: Attachments, Crossings, and Ethics*. Princeton: Princeton University Press, p 5.
5. Ibid. .
6. Kenneth Slessor, 1939. *Five Bells: XX Poems*. Sydney: F.C. Johnson.
7. Charles Baudelaire, 1964 [1857]. *The Flowers of Evil (Les Fleurs du Mal)*. New York: Bantam Book, p 185.

8

Cold Lazarus: Death in the Liveable City (Part II)

Now that we have our death goggles on, our hooded friend seems to be all around us. Despite all the chorusing of growth and endless abundance in free market Melbourne, the marks and totems of finitude suffuse the city for those that want to see. We want to see. We want this because denial of death is surely one of the worst neuroses that necrocapitalism visits on us, and the infliction feels heavy in a city dedicated to freewheeling growth. We tramps want to restore ourselves by finding death. Death cannot stop the tramps in their tracks, indeed its eternal grasp should bind them to the task of exposing the ordinary, if denied, maladies of the industrial city.

We think of poor Goethe in his eightieth year, hearing of the death in Rome of his hapless son August. Would such a painfully intimate death close down the mind journey of this most (if we may) august tramp of modernity? Not a bit of it. Goethe's famous utterance on hearing the sad news was (roughly and variously translated): *Over the graves then, onwards!* (He kept working on a signature work, *Faust*, until his own death a year later.) We take this as a great injunction, a call from the grave to move forwards in our city journeying, with respect for death

Courtesy of Michael Leunig © (http://www.leunig.com.au/).

but without the curious mixture of denial and sentimentality that has attended it in modern capitalism.

The cemeteries were a good start for the death tramp. But what about the ordinary fabric of the expanding, intensifying, bustling city that surrounds us? Where is our bony-fingered hooded friend to be found here? A rather striking and very new instance comes into view as we wander from our university, southwards on Swanston Street—which reputedly carries the busiest tram line in the world—towards the city centre.

There has been much redevelopment of the older built fabric of the street in recent years, partly driven by a private accommodation sector that caters for the booming numbers of foreign students that seek quarters in this neighbourhood, between the education monoliths, the University of Melbourne, and RMIT University. As we stroll south towards RMIT University, which sits on the inner-city edge, we encounter a new urban ghoul that has arisen like a corpse phoenix from a corner block development. We observe that this spectre seems part of a new urban species; another example will shortly follow. The Urbanest Student Accommodation development has transformed a previously lethargic set of corner sites into a massive residential offering for students. Its glassed, soaring bric-a-brac bulk looms over the ever-flowing traffic river that is Swanston Street.

Brendan Gleeson ©

Urbanest is physically and financially a ziggurat of development ambition arising on the coattails of contemporary Australian higher education, the nation's third most valuable export 'industry'. Before-COVID, at least, over 600,000 foreign students, mostly from Asia, provided fee income that props up a publicly underfunded tertiary education sector.

They also pay rent; a stream of value for a booming private accommodation sector. Urbanest competes with the equally starkly named 'Unilodge', and many others, for a slice of the valuable foreign student rent cake.

So why are the tramps distracted by the sight of this new student ziggurat, just one amongst a proliferating many? It is the shock of the new, to quote the late Robert Hughes, who had an eye for the tricks of modernist capitalism, that we otherwise know (in code) as innovation. There is something disturbingly innovative here… a strange reticence of the whole thrusting development which amazingly makes way for a historical building on the site corner. The towering Urbanest pyramid concedes valuable ground on its very street corner front to a meticulously restored historical building, etched in its graceful Victorian frontage as 'The English Australian and Scottish Bank Ltd'. It recalls an important colonial (and later) finance institution that has passed into history, absorbed in 1970 by ANZ, one of the Australian banking titans that exist today.

We tramps recall this corner building for many years as a mouldering commercial, sometimes retail, site, that looked like much of the rest of Melbourne's many neglected, often eviscerated, historical buildings. The Urbanest redevelopment led not to the all too common outcome for a historic building, its destruction and erasure, but rather its meticulous restoration, and most amazingly a spatial concession from the surrounding new building bulk that allowed it to continue to exist. As Urbanest arose on the graves of former sites, the little historic structure on the corner was restored to its original Victorian glory. Money was clearly spent—a century perhaps of neglectful, disrespectful reuse of the structure was reversed by the new development that exhumed and prettied up its corpse. Its original bank purpose had been long lost to layers of reuse and repaint but was now honoured with a restored corner head sign that (re)announces 'The English Australian and Scottish Bank Ltd'. Curiously the tramps see this more as an epitaph than living memory. The painted corpse is now an 'EzyMart' convenience store, retailing expensive provisions for impecunious students and others. There's life in the old girl yet, it seems.

8 Cold Lazarus: Death in the Liveable City (Part II)

We are disturbed by Urbanest and its enveloping accommodation of this pesky little historic building. What explains this unusual 'development outcome', we cannot say. The development-regulatory world is infernal and bent to the evolving interests of capital not to public interest, let alone historic sentiment. The tramps cannot imagine what produces this twist in development practice. We are, however, enchanted by a new turn that sees history unearthed through restoration not interred through demolition.

Until recently, historic urban fabric was open season for developers, especially those areas not protected by the city's weak and partial heritage controls. Melbourne, like many freewheeling growth cities in the world—Global North and South—has been steadily erasing and brutalising its historic built legacy. The city has been in recent years lauded for its liveability and its recent design excellence, and can reasonably claim that some of this is true. But equally, as noted in a previous chapter, it can be seen as a 'Devil's Playground' for developers whose lust has been conjugated by public officials, eager to deregulate urban development in the interests of growth. In Melbourne, the responsibility for much of this urban damage lies with the state government, not the city of Melbourne. The latter's decision-making powers in planning were long ago removed and assumed by a State administration too eager to see cranes on the skyline as the glad tidings of growth.

What disturbs tramping sense at Urbanest is the accommodation to built history, a departure in development practice which commonly saw erasure of heritage. The graceful 'Marvellous Melbourne' legacy of the nineteenth century has been steadily abased and removed by development ambition since the 1950s. What to make of this sudden restorative urge? The tramps scouring Melbourne see growing instances of this 'restore don't bury the little corpse' approach to major redevelopment. The rescued historic cadaver is always dwarfed by the physical size and the budget of the enveloping development. They are codependencies; the prettied-up corpses of yesteryear and the brash McMansions of the now. A business model binds them, in life and death.

Ah that's it. The code is cracked. We recognise the infernal embrace of Eros and Thanatos in this 'exciting new field' of contemporary development practice. We can read the code but can't discern its authorship.

We suspect an overdetermination perhaps of weak regulatory pleading ('please be nice to historic buildings') meeting the always shifting (and authoritative) edge of entrepreneurial ambition (perhaps now, 'a little bit of history sells well…Old Melbourne for the student experience'). We speculate.

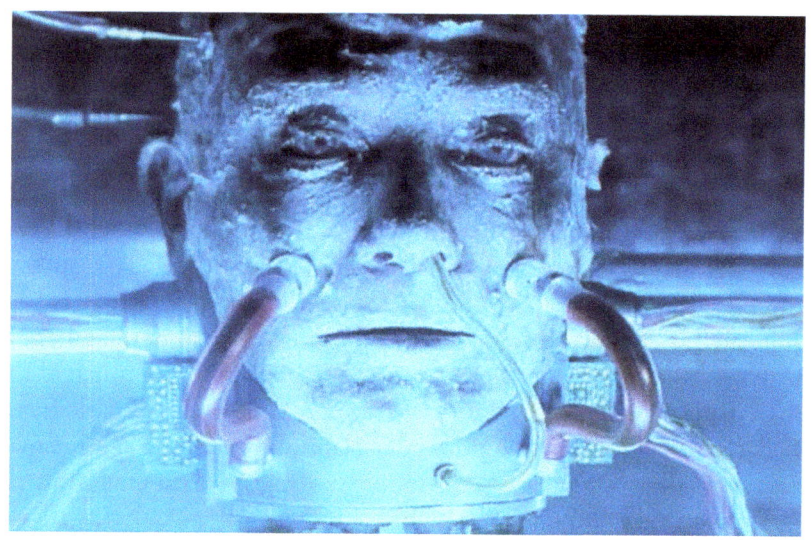

Courtesy of Whistling Gypsy Productions/Channel 4 ©

Amidst this ordinary confoundment, a clarifying memory suddenly appears—most appropriately from the landscape of the dead. What comes to mind is a gripping, grasping image from *Cold Lazarus*, the final television series of the English dramatist and screenwriter Dennis Potter. Potter was dying as he wrote the script, making time for a memorable broadcast interview in 1994 with critic Melvyn Bragg. He was an avowed foe of mega capitalists, most especially Rupert Murdoch—he told Bragg his cancer was named Rupert. Potter dedicated his last days to writing a script for the future; a story as it turned out of his pen about the radical, not to say vile, intrusions on human selfhood that was threatened by globalised capitalism with all of its ambitions to extract ever more value from a declining natural residuum.

Cue *Cold Lazarus*. The series was made after Potter's death jointly (at his insistence) by Channel 4 and the BBC. What interests the tramps is the core horror of its narrative, somehow recalled as we behold Urbanest today. The story, in short form, is set in twenty-fourth-century Britain, a scripted dystopia we do not need to describe now. Its main part for us is the reanimation of the head of a human being, Daniel Feeld, from our era in a lab dedicated to medical advancement but overtaken and manipulated by a global media oligarch. Feeld has donated his body at point of death to medical progress, and now centuries later a fragment (his head) is desired by a media baron who sees value in retailing the history of its memories, faithfully extracted by a sponsored medical laboratory.

Cold Lazarus is a complex, historically prescient tale written by a dying man. Potter was public about his death in a culture that denies the idea of ending, let alone dying. Readers should watch *Cold Lazarus* (widely accessible) to get a fuller appreciation of what we must here reduce to one image and compelling trope. In *Cold Lazarus*, Feeld, as a preserved laboratory head, slowly wakes through the course of a wider roiling narrative to some form of consciousness. We eventually learn that he awakens to know he is dead. We learn that the only wish of his reanimated head is to be allowed to die. His reawakening vestigial humanity recoils at what he experienced his entire waking life; the value extraction of all things living and not by an ever-evolving capitalism.

We look on Urbanest and the preserved and prettied corpse of 'The English Australian and Scottish Bank Ltd'. Its cries for death move us. We are at once charmed and haunted. Our guiding philosopher maintains that 'enchantment is something that we encounter, that hits us'.[1] We consider ourselves struck.

Note

1. Jane Bennett, 2001. *The Enchantment of Modern Life: Attachments, Crossings, and Ethics*. Princeton: Princeton University Press, p 4.

9

Walking the Corridors of Consumption

The atmosphere is defined by that bright, fluorescent, white light, almost like that of a petrol station or a surgery, exceedingly unnatural. As if our imperfections were not noticeable enough, here they are starkly on display. People are hurriedly moving in all directions, barely avoiding collisions, shopping bags in hand. We think we saw one consumer frothing at the mouth—whether they were suffering from affluenza or rabies, it is hard to say. Twisted faces glance at twisted faces, ours included. Something isn't right, but what is wrong is not entirely clear. We are the people of late capitalism, together, alone.

Against the background of chatter and clatter, a sterile, formulaic pop song is audible, emanating from some indistinguishable location. Somewhere milk is being steamed for a take-away latte. Someone tries to hand us a leaflet about a 'revolutionary' new phone deal. A toddler is having an impressive tantrum. It is all a sensory overload, especially when advertisers are everywhere competing for our attention. In quiet contrast, an old Italian couple hobble slowly past, eyes sensibly cast downward, with arms linked in mutual support, in a world much changed. Perhaps they are as lost as the rest of us. How did we get here? And how in hell do we get out?

Courtesy of Lena Singla © (http://lenasingla.com/).

Exactly why, we are not sure, but we've found ourselves wandering a shopping centre in Melbourne's northern suburb of Campbellfield—although we could almost be anywhere in the industrialised West. If we were teenagers we'd be at risk of being arrested for 'loitering'—that crime of remaining in a particular public place without any apparent purpose. Who hasn't been guilty of that? Unlike us, however, it seems the patrons of this vast plaza are brimming with purpose and intent. Huge glass windows line these corridors of consumption, not to allow light in but to capture the gaze of the unfulfilled consumer. Desire must be created—and we feel the system beginning to work on us as we become self-aware of how tired our clothes look. This morning they were not a cause for any discontent. Those glossy ads just made us poorer, damn them.

Products are carefully on display, as if they were the Crown Jewels, except everything here is on sale. Up to 50% off, just for this weekend—but conditions apply. Conditions always apply. But we're never told if the products were made in a sweatshop. The consumerist aesthetic is so slick and meticulously choreographed, but what violence does this plaza hide from its patrons? What do the patrons hide from themselves and each other? And who are we to be asking these questions? Everyone is searching for meaning after all, and some happen to search for it in shopping centres. In some sense, the search for meaning is why we are here too. That shirt sure looks nice.

Courtesy of Lena Singla © (http://lenasingla.com/).

In his short but provocative article 'The Magic of the Mall',[1] Stephen Horton writes of how the desire of the consumer is a guilty one, like the desire of the man who wanders the red-light district of Amsterdam, surveying the merchandise, examining the things-for-sale in the windows, seeking the passive gratification of consumption without reciprocity. 'The guilt of the mall shopper is no different', Horton explains. 'The mall is a domain of fashion, the home of the brand name. The value of its merchandise lies not in use but in the status bestowed by possession. Such merchandise is part of the modern cycle of consumer dependency. The purchase of one status symbol fuels the desire for the next. Guilt is desire with no end'.[2]

Ecological economist Tim Jackson makes a similar point in his book *Prosperity Without Growth*: 'It is precisely because material goods are flawed but somehow plausible proxies for our dreams and aspirations that consumer culture seems on the surface to work so well. Consumer goods… provide us with a tangible bridge to our highest ideals.'[3] Jackson adds that they fail, of course, to provide genuine access to those ideals, 'but in failing they leave open the need for future bridges and so stimulate appetite for more goods'. Consumer culture perpetuates itself, he concludes, 'precisely because it succeeds so well at failure!'[4]

As we look around this shopping mall, we see that success starkly on display. Most troubling is the empty gaze, not of the consumer, but of the shop attendant. Does any divinity stir within them? Like robots, they stand still, staring blankly into the middle distance, until, at last, a prospective purchaser enters the store. The robot jolts to life, forces a smile, and pretends to care deeply about your day and desire. Their attention is effusive, almost oppressive, but always subtly encouraging, especially when the store manager is watching. You look great in that. A shirt to match? How about a scarf? The colours are perfect complements. These shoes are on sale if you buy two pairs. The subtext is: life is not sufficient as it is. You need to buy something to make life better. But is it really more stuff that is lacking? To paraphrase Thoreau: people go fishing all their lives, not knowing it is not fish they are after.

In wandering these corridors of consumption we wonder. It is hard to know what is true and what is false in the social landscape of a shopping mall, since the more that human relations are mediated by the profit motive, the less real those relations become; the less we can trust them or take them at face value. There is always an ulterior motive. After all, if shop attendants knew we weren't going to buy anything, would there be a pleasant greeting or grateful farewell? Probably not. Subconsciously, then,

9 Walking the Corridors of Consumption

Courtesy of Julie Reason © (Warmred Studio).

we learn not to trust each other, assuming people only express interest in others if they sense an opportunity for a sale.

The problem is that, outside the mall, when we are lucky enough to receive a *genuine* greeting or farewell, we may miss the sincerity, miss the opportunity for real connection, mistaking it for the false interest of the store attendant. Capitalism thus leaks well beyond the marketplace, reaching the depths of human society and reshaping life according to the universalising logic of profit-maximisation. Increasingly it is hard to find space 'outside' capitalism. But still, we must seek and defend those temporary autonomous zones, should we manage to find or create them.

Courtesy of Greg Foyster © (http://gregfoyster.com/).

Back inside the mall, when the consumer leaves the store, the robot returns to their default state of bored, alienated existence. To be sure, the scholar, in what is perceived as an ivory tower, is not immune to this malaise, so we comment with empathy not in judgement. Fortunately, the wages earned can be used to resist capitalism, or so we might think. Through consumption, the defiant consumer searches for meaning that has been taken from them in their productive, working lives. What a devious and deceptive feedback loop. Again, it succeeds because it fails. Less meaning in life—more 'consumer resistance'. And so it goes on.

Here, for the first time, we do not so much offer a critique as raise concerns with Bennett's take on enchantment. 'Can commodities enchant?' she asks herself—and answers 'yes'.[5] She celebrates 'vibrant matter'[6] and even an 'enchanted materialism',[7] and she does so, we acknowledge, with considerable philosophical sophistication and insight. Who has not experienced the exciting (albeit temporary) 'consumer buzz', as one returns home with a new gadget or garment? Especially when framed by advertisements—encoded messages crafted by the world's most sophisticated psychologists to insidiously manipulate—there can be no doubt that material culture can, in fact, enchant. Evidently, there is little in the world that is more seductive; little that has sharper hooks.

Nevertheless, in today's consumptive and overly productive industrial cities, which are so violently demanding more from their ecological foundations than nature can sustainably provide, the task of *homo urbanis*, we argue, is to be very cautious, and to presumptively resist, the strong, alluring appeal of material commodities and artifacts. What is needed, in the words of critical theorist Herbert Marcuse, is a 'Great Refusal'.[8] Let us seek the good life elsewhere, in the diverse non-materialist sources of meaning, happiness, and fulfillment—privileging social relations, a saunter through the city, or low-impact creative activity over human-commodity relations. There are infinite forms of flourishing that do not depend on luxury and abundance.

Far from implying sacrifice or hardship, philosopher Kate Soper would call such post-consumerist orientations 'alternative hedonism',[9] implying that downshifted lifestyles of material consumption involve an exchange of superfluous stuff for more time and freedom that ultimately (or immediately) proves to be a good exchange, both for the practitioner of voluntary simplicity and the planet.[10] More significantly, perhaps, post-consumerist social movements would seem to be essential for creating the cultural prerequisites for any deep shift in political economy away from growthism.[11]

To be fair, Bennett is far from dismissive or unaware of the dangers of commodity culture. She endorses the environmentalist critique of consumerism and supports (without unpacking) political efforts to reconstruct the existing economic infrastructure which so often locks

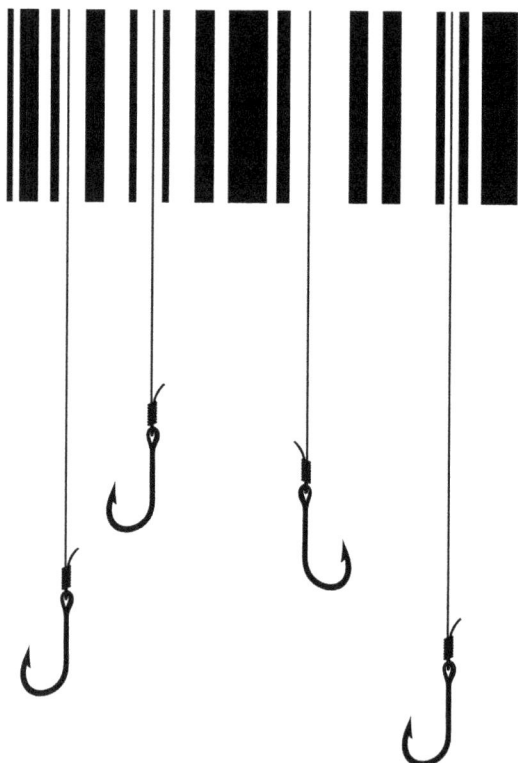

Courtesy of Chaz Maviyane-Davies © (http://www.maviyane.com/).

people into high-impact living. But when she says 'a modified organization of commodification and advertising could respond to the structural injustices in existing patterns of consumption without seeking to eliminate the enchanting effects of commodities',[12] we worry that she understates the risks of consumerist enchantment or the depth of structural changes needed to adequately address the injustices she rightly, but vaguely, acknowledges.[13]

Sure, material culture can be vibrant and enchanting. But in an age increasingly wrecked and wronged by the extractivism and waste streams seemingly bound to the material and energetic foundations of consumerist societies, we feel Bennett's balance of emphasis is a bit off.

Resolving the environmental burden is not a reformist project, as she implies. The system of globalised, consumer capitalism does not merely need 'modification'. What is needed is a new system of production and culture of consumption, which is a project, at least, of *revolutionary reform*. 'System change, not climate change', as the activist slogan goes.

All this raises a point deserving of clear acknowledgement: enchantment can be an ambivalent affect with ambivalent effects. Like a technology, it can serve various values and goals, such that *in itself* it is neither good nor bad. Rather, it is dependent on what cause it serves. In a world too often characterised and experienced as disenchanted, our presumption has been that enchantment is a mood that has the potential to evoke care and engagement, a spirit of generosity. We must accept, however, that it could also induce narcissism, individualism, or an inward-looking aestheticism that is blind to matters of suffering and injustice, both human and non-human forms. If the 'affective propulsions' of enchantment are to serve people and planet, we argue that *homo urbanis* should presumptively seek enchantment beyond the realm of consumer culture, not within it. This is obvious, but needs to be said.

From within Horton's 'Magic of the Mall' we read from the *Illustrated Guide to Paris*, published in 1852: 'The arcade, a rather recent invention of industrial luxury… is a city, even a world, in miniature'.[14] That is true, although we question whether today's gargantuan megamalls should ever be described as miniature. They are the industrial churches of capitalism, to which we are all expected to pay our tithes. This cathedral through which we tramp must cast quite a shadow over the clone-like Mainstreet outside, made up of a McDonald's, a Kmart, a Donut King, a Burger King, a Toy Kingdom, etc., etc. But inside, patrons are bathed in that sharp, white light, although there are no windows, just like in a casino. It is neither late nor early in the mall, time just is. Nature has no place in this shopping centre either, although everywhere there are plastic plants trying to put our primordial natures at ease. But they fail. A mild panic attack is simmering. Moving stairs are everywhere—for our shopping convenience—although none take us to the exit. We try to find the way out, with Radiohead's song 'Fake Plastic Trees' resonating in mind:

Courtesy of Pawel Kuczynski © (http://pawelkuczynski.com/).

But I can't help the feeling
I could blow through the ceiling
If I just turn and run

And it wears me out
It wears me out
It wears me out
It wears me out

Notes

1. Stephen Horton, 1998. 'The Magic of the Mall'. *Arena Magazine* (April–May), p 31.
2. Ibid.
3. Tim Jackson, 2009. *Prosperity Without Growth: Economics for a Finite Planet*. London: Earthscan, p 100.
4. Ibid.
5. Jane Bennett, 2001. *The Enchantment of Modern Life: Attachments, Crossings, and Ethics*. Princeton: Princeton University Press, p 15
6. Jane Bennett, 2010. *Vibrant Matter: Political Ecology of Things*. Durham: Duke University Press.
7. Bennett, *Enchantment*, p 156.
8. Herbert Marcuse, 2002. *One Dimensional Man: Studies in the Ideology of Advanced Industrial Society*. London: Routledge, p 74.
9. Kate Soper, 2008. 'Alternative Hedonism, Cultural Theory and the Role of Aesthetic Revisioning'. *Cultural Studies* 22(5), p 567.
10. Samuel Alexander (ed.), 2009. *Voluntary Simplicity: The Poetic Alternative to Consumer Culture*. Whanganui: Stead and Daughters.
11. Samuel Alexander and Brendan Gleeson, 2019. 'Degrowth "from Below": The Role of Urban Social Movements in a Post-capitalist Transition.' MSSI Research Paper, March 2019, pp 1–25.
12. Bennett, *Enchantment*, p 114.
13. For our extended statement on this matter, see Samuel Alexander and Brendan Gleeson, 2019. *Degrowth in the Suburbs: A Radical Urban Imaginary*. Singapore: Palgrave.
14. See Horton, 'Magic of the Mall', p 31.

10

A Riverside Ramble to the Last Hotel: Lostworlders Welcome

It is time these not-so-young men headed West. Every city has a western part—in Australian metropolitan parlance, 'the West' has so often been a metonym for the urban heartlands of the working classes. In Melbourne, Sydney, and Brisbane, this was historically the landscape of industry and blue-collar suburbia. It is an important destination for the tramps in quest for disruptive insights in the industrial city. We set out for its historical Melbourne heartland, the proud old suburb of Footscray. We trek in homage to the West and in particular one humble special place that we have heard about recently in national media, 'The Last Hotel'—of which more soon.

Our tramp takes us northwest from the city centre along the arterial spine of a sprawling industrial latifundium dedicated mostly to the ever-busy work of freight handling and transport infrastructure. We journey beside a fast-flowing stream of industrial traffic, lorries roaring and grinding, interspersed with tentatively driven human transports (cars and some very brave cyclists)—all surrounded and soundscaped by the noise of industry. A leisurely if somewhat deafening gambol of about 6 kms brings us to the edge of freightland, and the crossing point to Footscray, the Maribyrnong River. Ah, this beautiful industrial survivor… in the

Courtesy of Michael Leunig © (http://www.leunig.com.au/).

great Australian tradition we shorten its name affectionately to 'the Bong'. We approach its smoky waters with good feeling.

We admire a river that still obviously flows and lives in spite of the usual urban history of its treatment as an industrial sewer since the nineteenth century. Rivers and smaller watercourses were the first drainage systems for the industrial city in Europe and its new worlds. This was a strategy that worked only briefly until their befoulment threatened the health of entire metropolitan populations—even capitalists and their politicians! 'Enlightened self-interest' of the bourgeoisie as much as true human benevolence drove the project of urban reform. Sanitary reform and town planning brought a measure of relief to cities such as Melbourne, which managed to guarantee a clean water supply before London did.

Nonetheless, major urban waterways continued to be treated as drains into the twentieth century and certainly off-limits to sane swimmers and fishers. And the treatment of this city's rivers remains casually reckless and uncaring. During the writing of this book, for example, Edgar's creek in North Coburg was turned bright pink for reasons that were at first mysterious, but—surprise, surprise—in time an 'industrial' explanation was arrived at: spilt dye. Another proof that urban experiences can be both enchanting and disturbing: a bright pink river!

Courtesy of EPA ©

Eventually attempts were made to clean rivers but it was globalisation and the exodus of dirty industry to the Global South that really set the scene for their partial renewal. So it was with the Bong, whose banks and environs hosted an array of successful and highly corrupting manufactories from the Victorian era until the arrival of globalism in the 1970s, and the subsequent fairly swift relocation of all industry, much of it clothing and footwear manufactories, to various parts of Asia. If, today, the rivers of our so-called 'post-industrial' city are cleaner, it is because we have outsourced our dirtiest manufacturing elsewhere. As if pollution out of sight were evidence of cleaner production.

The Maribyrnong is the river of the West, wending its way some 40 kms southwards from its headwater region in central Victoria to its confluence with the city's other major waterway, the Yarra River. The Bong's lower reaches are tidal to the sea and it gets rather muddy and salty before it hits the Yarra. The first European naming of both rivers by invader-settlers in the early stages of the nineteenth century is starkly suggestive of their natural ecology as well as their later, post-settlement social ontology. There was both accuracy and historical consequence to these original settler brandings. What is known today as the Yarra River was baptised as the Freshwater River and the Maribyrnong as the Saltwater River. Both rivers were to suffer the violence of Victorian industrialism, acting as drains for manufactories and every other urban activity. But over time it was the drinkable Freshwater (later Yarra) and its geographic dependencies that hosted the more salubrious residential and commercial urban development while the environs of Saltwater were assigned to the decidedly more brackish waters of industrialism.

The Bong's pleasant environs have hosted European settlement since about 1835, the founding of Melbourne itself as a colonial city. Its name recalls, however, a vastly longer human recognition and occupation, that of the Wurundjeri people, whose ancient lands comprise much of the inner western, the northern, and the eastern parts of contemporary Melbourne. As we tramp towards the bridge over the Bong that will deposit your travellers in Footscray, our thoughts search in wonderment for the 40,000 years of human occupation of this riverland. Melbourne remains an industrial city that whirs to the supposedly timeless regimen of the growth economy, but which is regularly and inconveniently dethroned by economic busts and human maladies. Yet compared to the deep time of Aboriginal presence this all seems like the faulty whirring of a stopwatch. By contrast, Indigenous time seems like a rock-solid sundial on human experience, recalling the millennia of our common pre-industrial heritage.

We cross the Bong's waters over a great splayed bridge of no obvious cultural or design provenance, and yet assuredly a crossing of long civic investment and significance. This is a key historical transition point into the West. It was long ringed on its banks by large manufactories, but now the scenes are different. The banks we peer down upon have been freshened by a few decades of modest municipal greening, the replanting of eucalypts and shrubs and the installation of various pleasant and useful amenities such as walkways and recreation areas. It offers a typically Australian scene of scruffy piecemeal but well-meaning waterway renewal by cash-strapped councils and very often voluntary and civic effort.

To the left and on the far bank, we spy another instance of charming scruffiness, a wonderfully optimistic river touring operation which clings to a modest river berth. A smallish 'toot toot' ferry, the 'Blackbird', bobs about its mooring, the scene framed by an implausibly cheerful sign promising 'River Cruises' on the dear old brown Bong. We catch our breath at this refusal to see dirt in beauty, indeed industrially imposed and legated muck and grime, and march onwards fortified by the experience of gentle human knavery. Long may it reign in our city.

Finishing the crossing brings us to new rising citadels that avow the very antithesis of this humble optimism. We confront the soaring, mostly incomplete, residential towers that are rising on the western banks of the Bong where once lay rambling manufactories that once made humble clothing. Industrial manufactory has been erased and is being replaced by industrial residentialism. These seven apartment towers, so typical of Melbourne's 'renewal' areas, are the new artefacts of the urban growth machine, in this case the industrial production and maintenance of residential and ancillary space, evoking the types of population and dwelling densities known to the Victorian city. Early housing and planning reform sought to disentangle and decant their forebears through, *inter alia*, suburbanisation, and in a great historical swing they have returned in glittering glassed and well-plumbed forms.

When sold and inhabited these towering apartment blocks will replace the production of things (manufacturing) that previously marked these sites with the reproduction of bodies and labour-power, largely for the contemporary service economy that in many ways is every bit as industrial in its rhythms and expectations as the old banging of metal and sewing of fabrics and leathers. As our friend and colleague Boris Frankel might observe, the hopes of the 1970s 'post-industrial utopians' are negated in such citadels of accumulation.[1] Just ask their worker-residents.

Brendan Gleeson ©

We proceed from the bridge into the West through the eerily half-constructed towerscape. We are moved by the testaments to industrial-scale accumulation signed by the construction companies on the sites. In particular, we are impressed with 'Growland' company for its starkly signed mission, emblazoned across half-constructed towers to our right as we tread tentatively under temporary canopies protecting (we hope) foot traffic as it passes by these volcanoes of newly constructed ambition. As we pass, we check its website on our phones, and see promise that Growland '…is a developer that delivers on its promises'. Under shadow of its constructions (and their falling debris), we dare not doubt it.

We press westwards on our pilgrim's progression, 'The Last Hotel'. About 130 steps on from the fringes of the new towerscapes we arrive at our intended destination, a dishevelled Victorian-era pub, the most remarkable survivor of the previous Victorian world here, the Footscray

Hotel. As we noted at the outset, this crumbledown artefact of early industrialism had flashed briefly through the national media some time ago, catching the interest and the compass of the tramps. It was explained and lauded as a little bastion of social refuge in an era and area under the thrall of the urban growth machine. We now stand before its sturdy scruffy exterior, corner door announcing via a cheap small, neon sign 'Open'. A tramp's welcome.

Brendan Gleeson ©

Inside we encounter 'another world' of warm darkened assemblage, of normal pub stuff, tables and chairs and the bar of course, but also the belongings piled up of who knows? Crusty quiet male 'regulars' sit quietly near the bar clutching pints and 'pots' (in Melbourne measure, a half pint) while watching the women's football on the overhead tellie. As we journey around the old pub's rooms and spaces—bar, lounge (replete with large glowering fish tank, and more bric-a-brac furniture), kitchen, and beer garden—there is a strong sense of having dropped out of the growth machine into a very different time-space machine, that of simultaneous social acceptance and refusal. Let us explain.

The media slices foretold of the 'characters' that inhabit and run this small space that seems to bridge the old and new worlds of industrialism, as we have explained it. There is the ex-boxer, who serves ploddingly but pleasantly we thirsty tramps at the bar, watched quietly by Sharon, the Bar Manager. The Bar, as we observed, hosts a congregation of the quiet and alone, needing the company of alike. In the 'Beer Garden', a smallish hard space where cigarette smoking survives (more industrialism), a few groups of younger resolute types are comfortably entrenched. They seem expressive of the youthful grungy band culture that adheres to this outpost of the old world.

We know from the media that the pub is a crumbling redoubt that gives refuge to the lost and the lonely in a sea of development ambition driven on by the compaction visions of green urbanism. It is the last hotel that will have them. In media newsreels, its remarkable owner, Brian Kramer, enthralled us with his gentle evocation of the pub as a hub of acceptance and social mixing in an area fast giving way under municipal benediction to the dictates of the new urban industrialism represented by the abutting rising towerscapes. A particular refuge is the few crumbling residential rooms above the pub that host people 'with stories' not value in the capitalist sense. These 'down and outs' are hosted but also valorised by Brian and staff. There is a sense that if you stay here you can quietly repair, but good (enough) behaviour is assumed.

One media segment celebrates the story of a fellow, Lindsay, otherwise homeless, accommodated in the pub, who from there turned his life around. Another resident story related is that of Johnny, ex-prisoner and heroin-addicted for thirty years, who finds peace, acceptance, and finally clean living in this marvellous machine of human not material growth. How was the long cycle of Lindsay's brokenness broken? At one point in the media segment, Johnny gives a possible clue: 'Brian accepted me'. We venture to say he felt accepted without the stamp of exchange value, his humanity valued over his broken labour-power.

To our great fortune, as we tramps take a beer, Brian appears and comes to us, without knowing us. This is the old hospitality to strangers, perhaps, that predates industrialism. A half hour or more of warm discussion follows. He is so proud and welcoming of the social mixing harboured by the pub, the grunge youth, the old folk without close society, the crusty regulars, the lostworlders upstairs in the hotel rooms, the battlers (as we say in Australia) serving behind the bar, and all comers besides. Brian has had so many offers to sell out to developers, but never has and never will, he avows, while there is blood in his veins. It is a ground-level politics of refusal and courage in the industrial city and we pay tribute to The Last Hotel and all who sail in her.

Note

1. Boris Frankel, 1987. *The Post-industrial Utopians*. Madison: University of Wisconsin Press.

11

Guardians of Gandolfo Gardens

'For some months past an agitation has been on foot with a view to securing the lease from the Railway department of three blocks of land adjoining the Moreland railway station for the purpose of establishing reserves, gardens and children's play grounds'.[1] So reads a passage in Melbourne's main newspaper, *The Age*, on 1 August 1911, describing a suburban mobilisation that fought for, and ultimately secured, a beautiful plot of land for public use and enjoyment. This was achieved due to what was described at the time as 'the indomitable energy and persistence' of local residents. The lesson: nothing good comes easy. We must fight for the city we envision and love.

What was to become of this public land? In a 1911 edition of the suburban magazine, *The Coburg Leader*, we read: 'Judging from the plans this reserve will be an extremely pretty place when the trees are grown to a height at all and doubtless the children of future generations will bless the forethought of those citizens who succeeded in securing for them a breathing space before the neighbourhood became too much built upon'. Soon after, we read that, 'On Saturday over fifty trees of which 200 had been secured were planted in the new reserve in ground which had been ploughed the instant the weather became at all suitable'.

© The Author(s) 2020
S. Alexander and B. Gleeson, *Urban Awakenings*,
https://doi.org/10.1007/978-981-15-7861-8_11

One can sense the prideful achievement regarding this new community project—a public amenity that became known as the Gandolfo Gardens. 'The trees, which have received special attention, are growing splendidly, and it is anticipated that there may soon be a public garden in Moreland [being the greater district within which Coburg falls], where children may spend a pleasant afternoon'.

What might have animated residents to fight for this land and cultivate it with such care? We find an answer again in *The Coburg Leader*, which also provides some insight into the nature of Coburg back in 1911. It deserves quoting at length:

> The strongest and most plausible plea put forward at the meeting of Moreland residents on Saturday evening, for the securing of the railway reserve at Moreland station as a recreation ground, was the almost total absence of places where the children could play, and where the mothers could sit and enjoy a pleasant rest, coupled with fresh air. At present, Coburg is what would be styled south of the Yarra, a very open suburb. To the east and to the west particularly the latter large and open tracts of land are to be seen, while in the north, well Coburg ends and the country pure and simple begins. In the southern portion more generally known as Moreland, however, there is no doubt that the vacant spaces are disappearing with great rapidity, and that as Senator Russell said it will be difficult to secure fresh air spaces before long.

To modern Melbournians, the idea of Coburg being the place where the city ends and the country 'pure and simple begins' is well nigh unimaginable. Today Coburg is considered a suburb of the 'inner north', with five train stations and ten full kilometres of suburban development before reaching the suburb of Upfield at the end of its train line... and even then the city does not end. There is more than another 10 kms of suburban development before reaching Craigieburn, and still the city's sprawling tentacles are reaching further by the day. Like everywhere, it seems, Coburg is under pressure to 'densify' and move towards the 'green compact city', converting evermore parks and open spaces into multi-story apartment blocks, shopping arcades, or otherwise developed out of existence in the name of urban progress. In the words of Cat Stevens:

We're changing day to day, but tell me, where will the children play?

Well you've cracked the sky, scrapers fill the air
But will you keep on building higher
'Til there's no more room up there?

Despite these concerns, in 1911 the active suburban community had been successful. The land had been secured, providing residents with space for playgrounds, meeting places, fresh air, and a place to plant trees which would provide habitat to birds, bats, and other wildlife. But even then, warnings were imparted: 'The matter however will not suffer to rest here. The ball once started rolling must be kept in motion…' This sentence, also from the *Coburg Leader*, wisely recognised that a public amenity, such as this community park, although intended to serve present and future generations, is never secured in perpetuity, but only so far as it can be defended from the encroachments of alternative development proposals and pressures. As philosopher John Dewey once wrote: 'Every generation must accomplish democracy over again for itself'.[2] His point was that, at each moment in history, citizens and nations inevitably face unique challenges and problems, so we should not assume the democratic institutions, practices, and public goods inherited from the past will be adequate for the conditions of today. Our ongoing political challenge, therefore, is to 'accomplish' democracy anew, every generation.

Fast forward to January 2020 and many of those trees, planted in 1911, remain, towering over the suburb of Coburg with majesty, casting much-needed shade and providing natural habitat for a suburb, like all suburbs, under pressure from the urban development process. Soaring gums lie to the west of the station, homes for possums, wattlebirds, lorikeets, magpies, and hordes of other native wildlife. In the seating area, the remains of an Indigenous Canoe Tree is memorialised. To the east of the station, magnificent palm trees and many other mature native trees and shrubs contribute further to neighbourhood amenity and the lungs of the planet.

But we read online and see in the streets: 'The gorgeous Gandolfo Gardens and surrounding trees at Moreland Station are under threat!'

Courtesy of Agim Sulaj © (http://www.agimsulaj.com/).

The Labor Government's marginal electorate-winning 'major infrastructure project' is the Level Crossing Removal Project (LXRP), including the crossing at Moreland Station, which the Gandolfo Gardens surround. The government claims this initiative, amongst other things, will make travel times faster for drivers by removing boom gates at railway crossings—even though everyone knows that if you make driving easier and faster by building more roads and minimising stops, more people will drive and drive more often, and the commute will end up being just as slow or slower.

That well-known paradox of urban transport planning makes the consequences of the government initiative all the harder to swallow. The current LXRP design allows 113 of the park's trees—92% of them, many of which were planted back in 1911—to be destroyed. Back then it was said that 'residents would not get gardens unless they fight for them', a comment which we find restated on the website of Australian Greens MP Tim Read, who is calling for community mobilisation to protect the gardens. And some of the locals seem ready and willing to fight, as evidenced by the 'Guardians of Gandolfo Gardens' Facebook group that helps organise the resistance. On 15 January, fifty protesters from the Upfield Corridor Coalition (UCC) stopped work on the $460 million Bell to Moreland Level Crossing Level Removal Project (LXRP) for the third day in a row.

The local rag reports:

> As contractors John Holland attempted to erect fencing around the park at 8am, protesters linked arms, circled the truck and prevented workers from unloading the fencing. Twenty police threatened to arrest the group that included elderly residents and young mothers. At 10am, workers again attempted to cordon off the park but again the protesters linked arms, sang and resisted police and LXRP requests to move on. A Construction, Forestry, Mining and Energy Union delegate declared the site unsafe and half the workers left. This was celebrated by the defiant locals.[3]

A community organising website testifies: 'They tried to attack Gandolfo Gardens, but we have managed to fight them off with our peaceful protesting methods'. In a deep and disturbing irony, a few days later the chainsaw massacre of the towering trees was deferred again as extreme heat in the early to mid-40s meant work was not safe. In a suburb already short of old trees and green spaces, it seemed all the more insane that governments would allow the removal of what mature trees still existed, trees which functioned to mitigate the urban heat island effect and provide habitat for a declining biodiversity. Add to this the fact that Australia, as we write, is literally on fire, with 11 million hectares of bushland having burnt down in the early weeks of this infernal summer. We are entitled to ask with Cat Stevens, not only, where will the children play, but also, where will the birds and animals with whom we share this urban space live?

Courtesy of Michael Leunig © (http://www.leunig.com.au/).

Nevertheless, despite the insanity of destroying over one hundred mature trees in order to make it easier for drivers to get about—neatly summarising the logic of industrial civilisation—the trees have, in fact, been murdered. The public protesting, in which we participated for a time—more to bear witness than out of any sense of hope—was unsuccessful. To hide the bodies, the developers promptly put the dead trees through an industrial-scale mulcher. It was literally a massacre. The once beautiful community park is now ugly and horrific, an industrial wasteland. The urban tramp may be condemned as wandering loafer, not doing enough 'productive work', and yet we wonder why it is that 'the speculator, shearing off those woods and making the earth bald before her time', is esteemed as an honourable citizen, 'as if a town had no interest in its forests but to cut them down'.[4]

What happens now to the wildlife—the birds, the bats, the possums—which lived in the trees? Neither the government nor the developers seem much to care; it is a necessary evil. Nothing to see here, just the normal operation of 'sustainable development'. As ecospiritualist Matthew Fox writes: 'Humankind has been involved in a gross desacralization of this planet, of the universe, and of our own souls for the last three hundred years. Here lies the origins of our ecological violence'.[5] This is a story of disenchantment that must be unsettled. How can we consecrate our cities? Where can the sacred be found?

We have a friend who once commented, a few years back, how rich the bird life was in Coburg. Sitting in the backyard she noted that where she lived—only one suburb towards the centre of the city, in Brunswick—she didn't often hear birds. How sad is that? Imagine not hearing birds. Is that the destiny for Coburg too—for *homo urbanis*—as the city as 'growth machine' continues to chew away at natural habitat in our cities? We are reminded of the conservation biologist Rachel Carson's seminal text *Silent Spring*, which brought attention to how industrial pesticides—DDT, in particular—were increasing profits at the same time as annihilating the birdlife. It is a book often described as birthing the modern environmental movement, praise that is simplistic perhaps but not entirely without its justification.

Have you, gentle reader, ever seen a flock of rainbow lorikeets bolt across a blue sky at dawn or dusk? It is one of the most captivating things you will ever see, especially when, lying on the grass in a suburban backyard or park, the sun is low in the sky and the technicolour feathers of the parrots' underbellies are illuminated, as if divine. Who needs consumer artefacts or Netflix when these divine creatures fly through the skies? For the time being at least, this is a common sight in Coburg. Rainbow lorikeets look like something you'd expect to see in the depths of the Amazon rainforest—parrots splashed with bright greens, blues, yellows, and reds—but here they are, in Coburg. Again, for the time being. In 2018, it was reported that in the decades since 1970, global populations of verebrate species (mammals, birds, fish, reptiles, and amphibians) have declined on average by 60%, painting a grim but now understandable picture of 'industrial progress'. What does the future hold? Talk of a Sixth Mass Extinction seems justified. Evidence is everywhere—the evidence of absence.[6]

Courtesy of Agim Sulaj © (http://www.agimsulaj.com/).

In February 2020, as we tramp around the fenced edges of the Gandolfo Gardens, the site of the recent chainsaw massacre, we wonder: Where are those rainbow lorikeets we love going to live now, with their homes, an inconvenience, recently mulched? The parrots either leave, because the real estate is too crowded and expensive, or they shuffle closer together in the remaining space. How many of the glowing lorikeets are left and lost, like the people of late capitalism, without a home or community to live out their essence? Through our urban densification and development, it seems we are forcing 'wildlife densification' too—more birds in ever fewer trees.

11 Guardians of Gandolfo Gardens

Courtesy of Michael Leunig © (http://www.leunig.com.au/).

Do we face the same fate as the birds of suburbia? Harrowing though it is to imagine, should we also expect a silent spring in coming years? What then of our economics of 'utility maximisation'? Utility for whom? Let us reflect on what we will tell our children when they ask about the rainbow lorikeet in the past tense. It is difficult to love and protect what one does not know. Urban disenchantment thus leads to apathy, resignation, and apolitics.

Back in 1963, Carson offered her blunt but painfully astute assessment of the human condition:

> Mankind has gone very far into an artificial world of his own creation. He has sought to insulate himself, in his cities of steel and concrete, from the realities of earth and water and the growing seed. Intoxicated with a sense of his own power, he seems to be going farther and farther into more experiments for the destruction of himself and his world.[7]

But just as Carson isolated the problem, so too did she hint at an alternative perspective, an alternative approach to existence, a radical ethic of wonder:

> ...the more clearly we can focus our attention on the wonders and realities of the Universe about us, the less taste we shall have for the destruction of our race [and, we would add, the broader community of life]. Wonder and humility are wholesome emotions, and they do not exist side by side with a lust for destruction.[8]

This is at the foundation of the urban politics of enchantment we are trying to understand, develop, and express in and through our urban tramps. We have already called for sleepers to wake. We now call on those sitting at their desks or computers to stand up, to walk outside, and to tramp through urban landscapes in ecstatic witness. Be enchanted—and be disturbed. Let the gentle art of urban tramping unsettle you, as it unsettles us. Let it denormalise the apocalypse.

And let the rainbow lorikeets give us all energy, for we need them to be exactly as they are, and they need us to be better.

Notes

1. Unless otherwise referenced, all historical material-quoted matter in this chapter can be found helpfully archived at the website of the Upfield Corridor Coalition, available here: https://upfieldcorridorcoalition.org/2019/07/19/residents-campaigned-for-gandolfo-gardens-over-100-years-ago-now-we-need-to-do-so-again/ (accessed 5 June 2020).
2. John Dewey, in Jo Ann Boydston (ed.). *The Later Works, 1925–1953: John Dewey, Volume 13*. Carbondale, IL: Southern Illinois University Press, p 299.
3. James Conlan, 2020. 'Residents Stand Firm to Protect the Gandolfo Gardens.' *Green Left* (16 January 2020).
4. Henry Thoreau, 'Walking'. In Carl Bode (ed.), 1982. *The Portable Thoreau*. New York: Penguin, p 633.
5. Matthew Fox, 'Creation Spirituality'. In Michael Tobais and Georgianne Cowan (eds.), 1996. *The Soul of Nature: Celebrating the Spirit of the Earth*. London: Penguin, p 207.

6. Jeff Tollefson, 2019. 'Humans are Driving One Million Species to Extinction'. *Nature* (6 May 2019); Elizabeth Kolbert, 2014. *The Sixth Extinction: An Unnatural History.* New York: Henry Holt.
7. Rachel Carson, 1963. From Carson's speech in acceptance of the National Book Award, 1963.
8. See Alina Bradford, 2018. 'Rachel Carson: Life, Discoveries, and Legacy'. *Live Science* (31 March 2018).

12

Tramping Against Extinction: Counter-Friction to the Machine

It is one thing to tramp the city or a rainforest alone; it is a different experience to tramp with a companion or two. Still, nothing quite compares to urban tramping down the middle of the road, in joyful solidarity, with a large crowd of fellow citizens. This can be especially uplifting when it is clear everyone believes, with the fire of democracy burning in their eyes, that a new world is possible; that capitalism is a gross failure of imagination; and that we can do much better than the world *as it is*. Stephen Graham writes that 'tramping is at first an act of rebellion; only afterwards do you get free from rebelliousness as Nature sweetens your mind'.[1] In cities increasingly bereft of trees, one tends to remain rebellious.

From discontent, hope and vision are born, and as that dialectic unfolds, otherwise inert citizens can be provided with the 'affective propulsions'[2] needed for political engagement. This testifies, we submit, to the political significance of enchantment, for if one is not enchanted by the prospects of an alternative imaginary, why struggle? Why resist? Why tramp together or alone? Disenchantment makes a citizenry servile and obedient, resigned and hopeless, the practical effect of which is to keep people off the streets.

That is a political problem that requires an affective intervention to be effective—a problem sorely neglected by most political economists. The only antidote to disenchantment, we contend, is enchantment. This does not mean achieving an *enchanted life*. It means seeking a life punctuated with or graced by *moments of enchantment*, and developing strategies—like urban tramping—to deliberately cultivate that political sensibility, which is grounded in an ethics of care and engagement.

Still, as John Holloway writes: 'It is easy to forget that the beginning is not the word, but the scream'.[3] Politics begins in the body not the mind. 'Faced with the mutilation of human lives by capitalism, a scream of sadness, a scream of horror, a scream of anger, a scream of refusal: NO'.[4] What Holloway rightly and simply implies here is that *what exists is not acceptable to feeling and thinking creatures*. We are entitled to demand more from this absurd universe, from our absurd city. And if we don't yet know *how* we are going to build the new, more humane, and more ecologically viable world(s), at least we know we'll be muddling our way through the urban politics of life *together*, in the streets.

The overall sensory effect of collective tramping, to borrow from Bennett, is 'a mood of fullness, plenitude, or liveliness, a sense of having had one's nerves or circulation or concentration powers tuned up or recharged – a shot in the arm, a fleeting return to childlike excitement about life'.[5] Albert Camus once wrote: 'I revolt, therefore we are'.[6] Who ever said there was no such thing as society!

Today's crowd happened to gather at Gandolfo Gardens. This familiar meeting ground highlights how community mobilisation in the past can feed off itself to inspire or facilitate further mobilisations, sending transformative ripples out into the urban democratic future, further than one might at first think possible. As we've just described, these gardens were secured over a century ago by local activism, and here we are, (sub)urbanites of the twenty-first century, mobilising in the city again, supported and enabled by this social space. As more of the 'urban commons' gets privatised in our cities, we need to be aware that this means less opportunity to gather to practise resistance and self-governance, a contraction of space for public assembly that is a problem for democracy.

Capitalism has increasingly become a problem for democracy.[7] Hypertrophic urbanism is a problem for democracy. Indeed, as we see the

12 Tramping Against Extinction: Counter-Friction to the Machine

undue influence of money in politics ever more clearly in Australia and abroad, our growing concern is that so-called representative democracy is a problem for democracy. This morning we are gathering to tramp collectively through the inner northern suburbs of Melbourne, down the busiest street, as part of the Extinction Rebellion protests.[8] Could this emergent movement be part of what philosopher of deconstruction Jacques Derrida called the democracy 'to come'?

Birthed in the UK in late 2018 but now a global movement of movements, Extinction Rebellion (also known as XR) is emerging as one of the most prominent faces of the environmental movement around the world today, including in Australia. Admittedly, it has been quiet during the coronavirus lockdown (which is soon to disrupt this book and the world), for obvious reasons. But rumour has it that the rebellion has been active with online organising, suggesting that if there is one thing a virus cannot kill, it is a rebellion.

Extinction Rebellion has three bold demands of government. The first is to tell the truth about the darkening ecology of our shared Anthropocene, by declaring a climate and ecological 'emergency'. Being honest and truthful doesn't sound so demanding... until one reads the science and thinks through the implications. The second demand is to decarbonise the economy and halt biodiversity loss by 2025. Let's be realistic, as the saying goes, and demand the impossible. The third demand is to establish citizens' assemblies—groups of ordinary people, selected like a jury would be selected—that do their own research and thinking and then inform public policies on key environmental issues. If the governments are committing or facilitating ecocide, then they are breaking the social contract, calling their legitimacy into serious question. In such times of democratic crisis, alternative institutions are required, and Extinction Rebellion is making a plausible and coherent proposal deserving of serious consideration.

What distinguishes this new mobilisation from most other manifestations of the environmental movement is that it is explicitly defending peaceful civil disobedience as an important and probably necessary part of the social and political strategy for achieving a just and sustainable world. For instance, today we will be collectively tramping down the middle of Sydney Road, a suburban artery running through Coburg to Brunswick. When we get to the busiest intersection, we will stage a 'die in'. This will involve people lying down, as if dead from climate change impacts, in the middle of the intersection where they are not 'supposed' to be. The goal is to shut down the intersection and disrupt the city. The hope is to raise awareness, mobilise a broader constituency of climate activists through exposure to this mass mobilisation, and ultimately provoke a deep and swift decarbonisation of society. These efforts are only a drop in the ocean, of course, but every drop counts. And the oceans are rising.

While there are many aspects of Extinction Rebellion that deserve critical analysis, perhaps it is the commitment to civil disobedience that is most controversial, since it is a strategy *designed to disrupt* business as usual, operating outside the usual democratic channels. Civil disobedience can be defined as public, non-violent, and conscientious breaches of law undertaken with the aim of bringing about change in laws or government policies. Right-wing shock jocks are quick to dismiss participants in Extinction Rebellion as merely consisting of 'troublemakers' and 'criminals' and 'career protesters'. But as we look around the multitude gathered today, we see mostly concerned mums and dads with their kids; a surprising amount of people in suits; lots of passionate young people; and generally a cross section of society increasingly concerned about climate breakdown.

Uncomfortable though it can make people feel, it is important for a society to understand the motivations for civil disobedience and evaluate the reasons given for practising this radical and disruptive strategy for societal change. Most people probably have deep reservations and concerns about individuals or groups deliberately breaking the law to advance their social, political, or environmental goals—and with good cause. This isn't to be taken lightly. Nevertheless, we all need to appreciate that many of the most significant cultural and political advances over the last century owe much to social movements that engaged in civil disobedience as a primary strategy. One might think especially of Gandhi and the independence movement from British rule; the suffragette movement; and the civil rights movement. These esteemed traditions raise the disconcerting question: Might future advances in society also demand civil disobedience?

12 Tramping Against Extinction: Counter-Friction to the Machine

Love it or hate, your tramps are of the view that Extinction Rebellion and movements like it are almost certainly going to grow in coming months and years as more people around the world become politically frustrated, angry, scared, and directly impacted by inaction in the face of today's overlapping ecological and humanitarian crises. We could call this the 'Rebellion Hypothesis', and it speaks of a new age of activism dawning.

Courtesy of Jacqui Brown ©

There is not even a need to make an argument that Extinction Rebellion *should* expand—a question we leave open for readers to determine on their own. Our hypothesis is simply that Extinction Rebellion and related movements *will* expand, as behavioural shifts in society (or psychological tipping points) are provoked by the ongoing deterioration of Earth systems and rising existential threats to the community of life. Put otherwise, we are suggesting that inaction has diminishing marginal returns, which makes social mobilisations for change more likely over time, given that the real and perceived cost/benefit analysis of the environmental predicament and ongoing injustices tilt in favour of collective action.

Courtesy of Benn Eden ©

At some point, no doubt germinating in urban centres,[9] a movement of movements may ignite in ways that currently our imaginations cannot even begin to grasp. Promisingly, social movements in history have had a tendency to surprise, which is a grounded source of hope. We are not sure whether forthcoming social and environmental rebellions will be able to *save* the world, but we feel they are destined to *change* the world as the world changes us. That is what social movements do. As urbanist Timothy Shortell writes: 'One of the most profound feelings of power comes from [participating] in mass actions in public space.... it is primarily in the movement of a large group of people in public space that one experiences social movement power as a material fact'.[10]

12 Tramping Against Extinction: Counter-Friction to the Machine

Marching down Sydney Road this afternoon, there is a palpable energy and excitement in the air. One small group carries an impressively large gong, which they club ceremoniously. Others hold erect bright banners declaring 'The Ocean is Rising and So Are We' or simply 'With Love and Rage'. It seems that there is a collective rumbling in the world today, a growing 'affect' for resistance and rebellion. There seems to be a growing anger and anxiety about the troubled future that is unfolding day by day, and a growing sense that, if governments are not going to act decisively in response to today's overlapping ecological and social crises, then ordinary people will have to be the driving force for change.

Courtesy of Jacqui Brown ©

Let us ask with Thoreau: Are we expected to resign our conscience to the legislator? Why have a conscience, if we are simply expected to uncritically affirm all acts of government? We must be human beings first and subjects of the state afterwards. As Thoreau argued, 'it is not desirable to cultivate a respect for law, so much as for the right',[11] and indeed, he insisted that respect for law can, at times, make us daily agents of injustice. In relation to his own time, Thoreau argued that one could not be associated with the US government without disgrace, for he could not recognise as *his* government what was also the *slave's* government. He concluded that if a government's law is of such a nature that it requires you to be the agent of injustice to another, then, he argued: break the law. 'Let your life be a counter-friction to stop the machine',[12] he declared. 'Cast your whole vote, not a strip of paper merely, but your whole influence'.[13] We reflect on how the Australian government is every moment losing some of its integrity.

Whatever your views on civil disobedience, it is clear the global situation today would have been much less appalling if informed, compassionate political action had been taken decades ago. Further inaction and avoidance will only produce more desperation from ordinary concerned members of society, who will be drawn to social movements like Extinction Rebellion. Political inaction is an act, and sometimes—like today—it will provoke a reaction.

So will Extinction Rebellion end up on the right side of history? That question is not for your tramps to answer. But as this global movement readies itself for increased action (after-COVID), what is important is that people avoid superficial, knee-jerk reactions. We need a sophisticated and engaged debate on the complexities shaping this moment in history.

For now, however, the multitude of urban trampers comes to rest. We have arrived at the designated intersection, outside the Brunswick Town Hall. A group known as the 'Red Rebels' are waiting silently on the corner to welcome us with their arms raised in a V. They are dressed in what looks like red velvet, including red headdresses that resemble turbans, with what appear to be long red dreadlocks of fabric attached. Their faces are painted white with black features, as if determined to haunt the civilisation that haunts us all.

12 Tramping Against Extinction: Counter-Friction to the Machine

What does the red stand for, we wonder? The blood of lost species? Rage? Love? Emotion? Heat? Empathy? All of those things and more? In unison, the Red Rebels slowly change pose, this time with the revolutionary gesture of one fist raised in the air. The police presence is huge and intimidating but for the time being they let us all be. People begin to lie down, closing down the intersection, and for a moment everything is quiet. Soft music plays, overlain with a speech by the one and only Greta Thunberg. We see tears fall on the city streets.

The scene is as enchanting as it is disturbing. Hundreds upon hundreds lying, as if dead, in the middle of the road. Of course, some commuters begin to look annoyed and start honking their horns. Even though it is a Saturday afternoon, one clever dick shouts: 'Get a job, hippies!' Many more onlookers look sympathetic, intrigued, or even moved by the collective action unfolding.

Courtesy of Melissa Davis ©

The will to social justice, Bennett argues, can be 'sustained by periodic bouts of being enamoured with existence, [for]... it is too hard to love a disenchanted world'.[14] This afternoon some people who have been sufficiently enchanted by the demonstration defect from their role of passive bystander and join the 'die-in' demonstration. They sense this disruption has a point. It is giving people energy to fight for a safer, better world. Somewhere loudspeakers switch on and some music begins to boom vibrantly. People begin to stand up and move about, and the demonstration of civil disobedience begins to take on the atmosphere of a festival or carnival. And for the time being, at least, no cars pass. The industrial city is still.

The revolution may not be televised—but one thing is sure, it will occur in the city… and there could well be both death and dancing.

Notes

1. Stephen Graham, 2019 [1926]. *The Gentle Art of Tramping*. London: Bloomsbury, p 61.
2. Jane Bennett, 2001. *The Enchantment of Modern Life: Attachments, Crossings, and Ethics*. Princeton: Princeton University Press, p 3.
3. John Holloway, 2010 (2nd ed.). *Change the World Without Taking Power*. London: Pluto Press, p 1.
4. Ibid.
5. Bennett, *Enchantment*, p 5.
6. Albert Camus, 1951. *L'Homme révolté*. Paris: Gallimard, p 36.
7. See Boris Frankel, 2020. *Capitalism vs Democracy? Rethinking Politics in an Age of Environmental Crisis*. Melbourne: Greenmeadows (in press).
8. See Rupert Read and Samuel Alexander, 2020. *Extinction Rebellion: Insights from the Inside*. Melbourne: Simplicity Institute.
9. David Harvey, 2011. *Rebel Cities: From the Right to the City to Urban Revolution*. London: Verso.
10. Timothy Shortell, 'Introduction: Walking as Urban Practice and Research Method'. In Evrick Brown and Timothy Shortell (eds.),2015. *Walking Cities: Quotidian as Urban Theory, Method, and Practice*. Philadelphia: Temple University Press, p 13.
11. Henry Thoreau, 'Civil Disobedience', in Carl Bode (ed.), 1982. *The Portable Thoreau*. New York: Penguin, p 111.
12. Ibid., p 120.
13. Ibid., p 122.
14. Bennett, *Enchantment*, p 12.

13

The Monumental Army that Marches on the Spot

Yikes, he's done it again! The weirdy beardy man just leapt out at an urban tramp making his way along Queensberry Street in the Victorian-era locality of North Melbourne. It's not the first time that this hirsute bronzed figure has appeared suddenly in this stroller's side view causing startlement and consternation. He waits ever for the passers-by with a loaded spring in his step. Impressive footwork for a statue.

Your tramp is always dislocated by this monumental moment, reminding him in every ambushed instance of the disapproving father of a past lover emerging suddenly as an angry ghost to reissue his censure as cold statuesque stare. His steely effigy grips a rolled-up document in his right hand, recalling the erstwhile father's unhappy habit of swotting flies with a coiled, weaponised newspaper. Neck hairs are always aroused, more disturbed than enchanted.

In 1904, the classical German sociologist Georg Simmel diagnosed modern urban life as injurious to mental health. Its intensity and surprises constantly bruised the headspaces of poor moderns, especially the hayseeds seeking asylum from rural idiocy in the industrial city. Out of the village frying pan and into the metropolitan fire! It is a significant and lamentable fact that mental health problems now plague highly

Brendan Gleeson ©

urbanised western populations like no other age. Disenchantment under the grinding neoliberal dispensation, with all its human discounts and injuries, exacts a heavy mental toll. Julia Kristeva, analyst of melancholia, confirms this contemporary diagnosis, observing, 'the periods that witness the downfall of political and religious idols, periods of crisis, are particularly favourable to black moods'.[1]

We tramps know this very much in this waylaid moment. The statuesque Father-Out-Law presents another and particular face in this pressing metropolitan crowd, triggering affects; guilt, fear, abjection…Our mental health is stressed for a moment, at least until the

13 The Monumental Army that Marches on the Spot

short further perambulation brings us safely to the nearby Town Hall pub, where griefs as much as thirsts can be eased.

Our startled moment raises the question of public statues in the industrial city. As with most metropolises, Melbourne has a dispersed congregation of statuary that is a much-ignored legacy of its Victorian past. Like 3D fossils, these monuments capture for civic gaze a deceased estate ennobled by the hard graft of stonecutters and, frequently, the generosity of public subscription. Victorian sculptural rapture lifted up a generally dreary multitude of white male functionaries whose only diversity was a scaling in status; major to minor, demigods to pettifogs, monarchs to mayors, horsed to standing.

Simmel spoke of the press of the living, but what about the demanding presence of the memorialised whose numbers grew as remorselessly as did the boundaries of the Victorian city? Like corpses escaped from graves, they stalked the cities and boroughs of their achievements; pretty and petty, fabulous and infernal, civic and mean. We say that honours systems have always betrayed a perverse ability to venerate the venal as much as the noble. So it is that Victorian monuments record the arrogations of power as much as the achievements of benevolence. In the contemporary city, this zombie army continues to exist quietly in the urban throng, marching on the spot, but occasionally leaping out at unsuspecting tramps.

And there is a geographic distribution of the stone-faced patrician class. Generally, the further from the urban core, the lower the status, which after a time becomes positively municipal. In the metropolitan centre, amidst temples of power such as parliaments, cathedrals, and courts, are arrayed the potentates: aristocrats, generals, judges, and the odd favoured lyricist. The high-born ghosts in Melbourne's parliamentary district include British military hero General Gordon 'of Khartoum' (died 1885), as well as colonial establishment poet Adam Lindsay Gordon (died 1870), who according to our friends at Monument Australia, 'won the Hunt Club Cup, the Metropolitan Steeplechase and the Selling Steeplechase at Flemington [still the main racecourse] all on one day in 1868'.[2]

As with the Batman memorial we visited earlier in 'Grave Matters' (Chapter 7), the Gordon monument—erected by public subscription in 1931—attempts more erasure of Indigenous history and ownership of Melbourne, and of the continent generally, carolling that 'He sang the first great songs/These lands can claim to be their own'. For at least 60,000 years before this Gordon showed up, Aboriginal songlines described, preserved, and celebrated the laws that bound these ancient peoples to their lands. The honking lie on the Gordon monument sharpens our tramping eye to the bad magic that so often still lingers over colonial statuary.

As we journey further afield into the early Victorian-era suburbs and beyond, those recalled in bronze and stone seem to diminish in stature and achievement to include minor civic officials and jurists. We encounter one major exception to this rule in the suburb of Brighton, to the south, where we shall venture a little later.

To return, in this startled North Melbourne moment, to chance a closer look at our censorious father. Consultation of the plaque by his side reveals testament to Henry Robert Bastow (1839–1920), architect, literally, of the public revolution that shook the State (then Colony) of Victoria with the passage of its *Education Act* in 1872. This was a remarkable moment in local human history when, in the midst of swelling sectarianism and cultural tension, the education of children was proclaimed to be free, secular, and compulsory. At the time of the legislation's promulgation, the Melbourne *Argus* newspaper enthused:

> For the first time in this colony, the young will now have an opportunity of acquiring the rudiments of education unmixed with the leaven of sectarianism, and every child, no matter what its parents' circumstances may be, will receive at the hands of the state that key which, rightly used, unlocks whole stores of knowledge, from whose ample treasures the patient and industrious may freely help themselves. If due effect be given to the compulsory clauses, none will grow up in that gross ignorance which is such a fruitful mother of crime, which fills our goals, and yearly robs honest industry of a large portion of its reward.[3]

13 The Monumental Army that Marches on the Spot   141

The colonists were by stripe pragmatists not purists and an accommodation of public and private endeavour was permitted to meet these fine objects. Religious and other private schools were allowed, while a strongly resourced public education sector was established and avowed.

Many religious and private schools were already extant. So, the first task was to produce the public education sphere. Melbourne, as capital of the colony, was centre stage to a massive building programme that saw the construction of 615 publicly run schools in five years. These finely built edifices, customised to local needs—large, middling, and small—were quickly and expertly arrayed through the metropolis and the Colony's rural hinterlands.

Following passage of the *Education Act*, the English-born Bastow was appointed Chief Architect and Surveyor for the new Department of Education, which was tasked with realising an instructional system that guaranteed access to free and universal education. He was remarkably successful in his task, overseeing and in many instances personally designing an impressive constellation of public schools, most of which continue to this day.

Bastow today stands (we say lurks) outside 'STATE SCHOOL N° 307', an establishment opened in 1882, marking the halfway point of his remarkable first construction effort. This beautiful set of buildings, skinned with London-style polychromatic diagonal brickwork, were in recent times revived from dereliction and decrepitude by a new architectural effort that, *inter alia*, necessitated the removal of 6 tonnes of pigeon waste. The old school was caringly revived as the 'Bastow Institute of Educational Leadership', a state-run facility that provides training for Victorian educators. We see no guano on Bastow. And of course, the point must be made that this is a statuesque Victorian, not just a Victorian statue.

Our Father was reincarnated with the school's restoration and set in his memorial place in November 2011. And there he stands, a modern installation, startling passers-by with his bronzed depiction of Victorian virtue. We wonder how the dead can still produce such lively disturbance. We must salute their capacity to cause wonderment in the industrial city, even to this day. In this case, is it because Bastow's monument reminds us of the quickly fading era of universalism, the progressive principle which held that access to basic goods and amenities was a fundamental right and virtue of citizenship? Neoliberalism with its mean-spirited targeting of public services and preference for private provision has set out to unpick and devalue the universalist principle. Bastow would not be pleased and surely would not suffer this foolishness—that newspaper still looks coiled for action!

At the same time, we ponder whether universalism should always bind us to the inflexible, frequently heartless, modes of mass education that prevailed through much of the industrial era. What would a truly post-industrial education look like? In what forms would a post-industrial youth grow and mature into a post-industrial adult citizenry? Often questions are more easily asked than answered.

Footnote to Father… the cold scold that confronted the tramp, and evoked memories of distal relationships, proves in historical testimony to be a man wedded to Victorian virtues, namely benevolence and modesty. The great, and indeed here monumental, Bastow was in life a leading Melbourne member of the Plymouth Brethren, The Society of Friends…in a word, Quakers. Quaker values, especially simplicity, explain much about his life and contribution. We read that after a distinguished and generous public career, Henry Bastow, '…retired from public life to central Victoria, built a simple home equipped with a meeting room for his Brethren fellows, and became an apple orchardist'.[4] Ah what pleasant news. The tramp now makes peace with Father, and indeed repents of his incautious reaction to this paternal witness of public virtue. He has returned in statue form to enchant us, not spook us, as a model of civic generosity and solidarity, deeply at odds with the radical individualism of our neoliberal age.

We move away from the recently reincarnated 'Saint' Bastow, back to the deceased estate of dubious virtue we know as Victorian statuary. For every Quaker in Victorian Melbourne (a rare species locally) there were many fakers, the pompous civic grandees who in truth were venal seekers of public benefit and recognition. They exist in multitude and are not hard to find.

Our trek takes us southwards into Melbourne's 'lower' suburban landscape, the locality of Brighton. Here, somewhat bizarrely, on the sidelines of the busy Nepean Highway is a statue that commemorates Sir Thomas

13 The Monumental Army that Marches on the Spot

Courtesy of Pawel Kuczynski © (http://pawelkuczynski.com/).

Bent (1838–1909), former mayor of the municipality of Brighton and later Premier of the State of Victoria (1904–his death). Our crowd-voiced friends at Wikipedia describe the marvellously and appositely named Bent as '...one of the most colourful and corrupt politicians in Victorian history'.[5] Bent was a land developer turned politician; after his parliamentary election, he gained an influence over railway alignments, government spending, and other public 'superpowers' only dreamed of by capitalist competitors. He died a rich man.

Much mirth amongst the common folk of Melbourne has always attended the roguish memory of 'Tommy Bent'. It's easy (and importantly healthful) to laugh, and much of Bent's chronicle invites rueful

Courtesy of Monument Australia © (www.monumentaustralia.org.au).

mirth. And yet, here the tramps must acknowledge the modern industrial dialectic: that Bent, in spite of many accesses to personal gain, was a straight arrow of modernist development, 'a committed advocate of public utilities, railways, roads, tramways and gasworks'.[6] His public leadership produced civic as well as private gain.

In our neoliberal age, as noted above, we have seen public goods increasingly whittled away by the logic of privatisation and austerity economics, a strategy pitched to the public as being in their interests ('it is more efficient') while elites and developers get richer as the poor remain as insecure as ever. Our cities and societies are being harmed by this hollowing of the public domain in ways our modernist statues would surely frown upon. As Karl Polyani argued long ago, a 'Great Transformation' is underway as ever more aspects of our lives are commodified under capitalism and brought within the governing rule of the market.[7] As crafted by the range of neoliberal ideologues in positions of power

13 The Monumental Army that Marches on the Spot

over recent decades, society has become a subset of the market, when the latter must serve the former if it is to be justified.

Even the 'progressive' notion of a universal basic income (or UBI), favourable in large segments of the Left today, still commits a society to a mode of social organisation governed fundamentally by market transactions and mechanisms, and financing such a system risks committing a society to the very economics of growth that must be opposed if there is any hope of avoiding ecosystemic collapse in the foreseeable future. Without presently being able to provide any form of defence, we feel the notion of 'universal basic services' is an approach that genuinely can be called progressive, a policy platform that would ensure everyone in society had their basic needs met as a foundational social service.[8] This post-capitalist politics would thereby loosen the iron grip of market rule under neoliberalism and create the structural context that would permit more of our lives to take place beyond the market and in pursuit of goals other than monetary or material gain. We feel our statuary friends, if their heads weren't so full of bronze, might have been open to this radical notion, dedicated as they were to a broadly accessible public service.

We stand by poor Bent, marooned on a ribbon of verge on the supercharged, snarling Nepean Highway that connects central Melbourne to the distant (some 100 kms) extensions of the metropolis. The old road was progressively expanded and improved from the 1950s, and in the 1970s, the old man (installed hereabouts in 1913) was literally sidelined to a new widening in the Brighton area. Since then he has looked on, forlornly, at the ever-busier flow of the highway, recalling without audience a happier era of public (and privately helpful) endeavour. We look closely at his steely imposture, noticing the left hand, held out in bronzed supplication. Oh, sculptor cynic, is this the hand of graft?

Henry and Thomas—recently and historically committed to public memory as statuary. You are part of a major, if rarely remarked, estate of the urban dead, recalled to civic service as monuments. May we henceforth look you all in the (cold) eye. As human artefacts of the industrial city, we salute you!

Notes

1. Julia Kristeva, 1989. *Black Sun: Depression and Melancholia*. New York: Columbia University Press, p 8.
2. See http://monumentaustralia.org.au/themes/people/arts/display/107210-adam-lindsay-gordon (accessed 5 June 2020).
3. See https://en.wikipedia.org/wiki/Education_Act_1872_(Victoria) (accessed 5 June 2020).
4. See https://en.wikipedia.org/wiki/Henry_R._Bastow (accessed 14 June 2020).
5. See http://monumentaustralia.org.au/display/30512-sir-thomas-bent (accessed 5 June 2020).
6. Ibid.
7. Karl Polyani, 2001 [1944]. *The Great Transformation: The Political and Economic Origins of our Times*. Boston: Beacon.
8. Anna Coote, PritikaKasliwal, and Andrew Percy. 2019. *Universal Basic Services, Theory and Practice: A Literature Review*. London: UCL Institute for Global Prosperity.

14

Sisyphus in the Suburbs: Pushing the Rock

Today we tramp the outer Melbourne suburb of Upfield. Why Upfield? Why not? The choice was arbitrary. On this yawning Saturday morning, we just hopped on the closest metropolitan train we could find and took it to the end of the industrial rainbow, as far north as it would take us, in search of, not of a pot of gold, but moments of urban awakening.

Yes, more committed urban tramps would have travelled by foot. Google Maps indicate that it would have only taken three hours and twenty-four minutes to walk there from our university campus; a decent walk, but not unmanageable. As we consider but ultimately pass up the journey by foot, we recall the stories of psychogeographer Will Self, who once walked to Heathrow airport from London, from where he flew into LA airport, and rather than getting a taxi or bus into the city, walked to Hollywood. We sense him shaking his head at us for using public transport rather than using our own feet. How dare we call ourselves urban tramps!

This weekend, however, in a busy life where the dust never seems to settle, we choose not to tramp *to* the urban fringe but *around* the urban fringe—a metropolitan edge that is ever receding, like our hairlines, only densifying not thinning out as it creeps backwards over time.

Courtesy of Thea T © (https://www.redbubble.com/people/tatefox/portfolio).

Upfield may represent the end of the train line but by no means is the city satisfied. Melbourne's growth machine has long since expanded, both vertically and horizontally, beyond the urban imaginations of yesterday's city planners. It is not done yet.

In any case, we should not dismiss the privilege of being able to use this train line. It was almost lost not so long ago. Cash-strapped state governments in the 1980s and 1990s tried to close it rather than invest in its modernisation. Indeed, we read in Melbourne's newspaper *The Age* that hand-operated boom gates were still in use along the Upfield line until the late 1990s, while governments pondered whether

14 Sisyphus in the Suburbs: Pushing the Rock

to close the line forever. Hand-operated boom gates! Another profession extinguished by automation, perhaps in this case, at least, for the best.

Governmental threats about closing a train line are difficult to oppose, given that the passive act of simply closing something means there is no bulldozer to obstruct with the bodies of committed activists. Nevertheless, as ever, activists in past decades got creative. They conspired to delay trams by lifting wheelchairs and bicycles onto the carriages with deliberate slowness, giving activists time to hand out leaflets of opposition to the passengers, before hopping off and doing the same thing when the next tram arrived. They were like armed train robbers in the old West, but instead of stealing things at gunpoint they were simply leaving ideas, dreams, alternative visions. And we all know, as do our governments, that ideas are far more dangerous than guns. All the more subversive for being peaceful engagements.

Owing to this hard work of social mobilisation and consciousness raising, eventually the prospect of closing the Upfield train line became too politically damaging for governments to consider and the policy idea was dropped. Nik Dow, one of the agitators of the time, noted: 'Sometimes in a democracy you win'.[1] As we hop off the train in Upfield, we pay homage to those urban activists—easily forgotten, like so many others—who have helped shape this city. There is an important lesson here: a city either washes over us or we actively shape it. Which will it be?

We leave the station and enter the streets of this outer suburb, taking a moment to look back along the train line to get a sense of the city's reach. The CBD is imperceptible in the distant haze, with countless suburbs unfolding between here and there—and more behind us. Notably, the historian Graeme Davison describes Australia as 'the world's first suburban nation'.[2] Urbanisation was the great wave that carried capitalism through the twentieth century to the shores of global preponderance, freighted with the models and machinery for mass consumption and the lifestyles that enacted it. Its principal model-machine was the suburb, functioning as a giant blotter to absorb middle-class aspiration.

Of course, suburbanisation was, like all capitalist development, an act of creative destruction—both vastly consumptive, especially of nature and the fossil fuels, but also productive, of new nature (people, human ingenuity and capacity, lived experience). We too often neglect the latter, which speaks to the latent capacity in the suburban landscape to keep creating and producing new forms of species improvement. As we wander this typical suburb of Upfield—extraordinarily ordinary in every respect—we ponder the potential of this dormant capacity and consider what the suburbs might yet become.

As we tramp forwards, we step backwards, attempting to gain a perspective on the bigger picture of this urban landscape. What is the meaning of suburbia today? Or rather, in suburbia, where and how is meaning found or created? Almost since the emergence of this form of built environment, social critics have expressed concerns about the work-and-spend cycle that suburbanites seem to fall into, consciously or unconsciously, willingly or otherwise, lost in the quest for material upshifting. As the defining expression of carboniferous consumerism, there is much to be critical of here in suburbia.

Nevertheless, we should not forget that suburbanisation was at first the escape route of Victorian middle classes from hellfire industrialism. The great suburban exodus was for a time, at least in the West, a journey of liberation joined by a proletariat that eagerly grasped the immediate fruits of industrial modernity. Suburbia was its vast material expression—an industrially produced landscape that offered the fundaments of a good life to the masses.

Contemporary anti-suburban commentary lacks this historical appreciation. It neglects the profound liberation that suburbanisation realised for a proletariat for whom the 'urban village' usually meant a crowded tenement without amenities or privacy. It would be equally wrong-headed to regard suburbia as a timeless mechanism for improvement. It was also a model of human growth freighted with the very same self-endangerment that threatens us now, but this was not to become clear until late into the twentieth century.

The sociologist John Urry reflected on the 'high carbon lives'[3] that were born and ordained in suburbia's car dependent fabric. As we have observed, the suburb represented and accomplished the dialectic of modern urbanism, creating *and* destroying human possibility. It was simultaneously a landscape of progress for many, including an improving

14 Sisyphus in the Suburbs: Pushing the Rock

working class *and* a central expression of the ecocidal process of overaccumulation. And yet, material progress or destruction aside, what of the existential condition of suburbanites today?

According to Greek mythology, Sisyphus was condemned by the gods to roll a rock up a mountain, only to watch it roll down the other side, and to repeat this futile labour, over and over again, for eternity. Urban activists today will empathise with this Sisyphean labour which so often offers no (immediate) rewards. Might this myth still hold some relevance for suburban life in the twenty-first century?

Courtesy of Julie Reason © (Warmred Studio)

In 1942, the philosopher and novelist Albert Camus wrote a philosophical treatise on this myth, using it as a metaphor to describe the absurdity of the human condition, declaring that the labours of human existence, just like the efforts of Sisyphus, were meaningless by any external or cosmological standard.[4] But rather than advocating suicide or nihilism as the solution to this human predicament, Camus defiantly embraced humanity's absurd fate and formed a philosophy of living in which the human subject was required to *create* meaning rather than pretend to find meaning in some unfounded metaphysical dogma. Thus, the objective meaninglessness of the human condition was not a cause for despair. Indeed, Camus closed his essay, rather obscurely, declaring that despite everything, the 'struggle itself is enough to fill a man's heart. One must imagine Sisyphus happy'.[5]

As we wander the quiet streets of Upfield, we accept Camus' invitation to reflect on the big questions of meaning and happiness, without hoping for answers. The myth of Sisyphus resonates especially in the light of our book's themes of urban development and the prospect of descent; of enchantment and disturbance; of creative destruction. Since industrialisation, humanity has been pushing the rock of economic growth up a vast energy mountain of fossil fuels, but the same rock which still absorbs and defines humanity's labours today has obscured the view of what lies ahead.

As the rock is pushed, people look backwards and assume that the future must look more or less like the recent past—a constant rise in energy supply and productive capacity—and so our species continues to march ever upward, not seeing that beyond the rock we approach the peak of energy availability—perhaps 'peak everything'. Like all mountains, this one too will have a bumpy downslope, but as Camus wrote: 'There is no sun without shadow, and it is essential to know the night'.[6] The dark night of COVID-19 will soon fall upon the world, but presently we tramps are blissfully ignorant, concerned 'merely' about such matters as our house of Australia being on fire.

Many people today seem to deny the prospect of civilisational descent and say that the existing forms of urban and suburban life are non-negotiable; that human ingenuity will constantly push back or transcend any environmental limits. But in fact, the laws of physics are non-negotiable, and our cities must and will abide—whether by design or disaster, that remains to be seen. From the perspective of deep time, the extraordinary but brief upslope of the industrial era will prove to be but a brief anomaly in the ongoing story of our species. The non-renewable nature of fossil fuels means that what goes up must come down. Today we look forth from atop this dark mountain, even as we tramp through the flat streets of Upfield.

We had come in search of urban awakening but were not so provoked. Today, at least, Upfield offers a mundane experience, so we tramp inward instead in search of enchantment. Taking refuge on the banks of Upfield's Merlynston Creek, overlooking the trickle of water through a concrete conduit, we ponder: Of what did Sisyphus dream as he pushed the rock? In that moment at the peak of the mount—the turning point—what did he see? What animated him to continue the struggle?

Journalist Rebecca Solnit notes, 'the future is dark, with a darkness as much of the womb as the grave'.[7] Nobody can know for sure how the urban future will unfold. Black swans might lie around every bend in the river (a line crafted before the shock of the virus struck globally). What, then, is to be done within these urban boundaries to intervene positively in these difficult, troubled, but uncertain times? More fundamentally, perhaps, how is one to *be*? Carbon civilisation has evolved over the last two centuries and produced a particular suburban form. This concrete jungle has atomised our cultures, decimated communities, all but abolished the necessary human connection with nature, and seemingly deadened the human spirit by fetishising technology and consumption.

And yet here we are, whether we like it not—like Sisyphus in the suburbs. Just as Camus was trying to understand the existential void left by an absent God, we find ourselves tramping through an age—an urban age—where the old dogmas of growth, material affluence, and technology are increasingly exposed as false idols. Like a fleet of ships that have been unmoored in a storm, *homo urbanis* is drifting in dangerous seas without a clear sense of direction. Where are the new sources of meaning and guidance that all societies need to fight off the ennui? Pioneering socialist Emile Durkheim used the term 'anomie' to refer to

Samuel Alexander © (http://samuelalexander.info/).

a condition in which a culture's traditional norms have broken down without new norms arising that are able to give sense to a changing world. Perhaps that is what best explains the urban condition today. We are coming to realise that we have lost our way, as the transformations that are supposed to represent 'progress' according to dominant cultural myths are increasingly experienced as breakdown.

Nevertheless, might this time of crisis actually shake us awake and provide an opportunity for some unexpected existential rejuvenation? We leave the reader to digest the inner dimension of these themes, noting only that the next and dangerous stage of our species' journey may require a reappraisal, not dismissal, of what we might loosely call 'spiritual' engagement with the world. This is not a call for any resort to dogma, of course. Camus would have none of that. We all have a rock to push, and in Upfield, Sisyphean rocks were lying around everywhere. Let the struggle fill our hearts.

Notes

1. Adam Carey, 2016. 'Not the End of the Line: How People Power Saved the Upfield Train Line'. *The Age* (5 June 2016).
2. Graeme Davison, 1995. 'Australia: The First Suburban Nation?' *Journal of Urban History* 22(1), pp 40–74.
3. John Urry, 2011. *Climate Change and Society*. Cambridge: Polity, pp 64–65.
4. Albert Camus, 2000. *The Myth of Sisyphus*. London: Penguin.
5. Ibid., p 111.
6. Ibid., p 110.
7. Rebecca Solnit, 2016 (3rd ed.). *Hope in the Dark: Untold Histories, Wild Possibilities*. Chicago: Haymarket Books, p 5.

Part III

AC (After-COVID)

15

Virtually Tramping Through Post-Normal Times

There is no way to begin this part of the book, at this time, other than by acknowledging the remarkable, mind-bending moment in which we write these words. It is a time of pandemic, one destined to shape the future of human civilisation for years, if not decades ahead. In Australia, the economy has all but shut down, with little open for business besides medical centres and hospitals, supermarkets and food outlets, and a very select number of other essential services. Against every ideological bone in its body, our conservative government has announced unprecedented stimulus packages, to avoid masses of people in our affluent nation from falling into destitution. The Federal government's 'jobs and growth' mantra now seems terribly outdated in these post-normal times, a quaint reminder of when the economic engine was turning.

Because so many people have lost their livelihoods, banks have had to freeze mortgage repayments for six months and rental evictions are currently prohibited. The national and state borders have been closed, and public gatherings of more than two people are banned. Someone in New South Wales recently was fined for eating a kebab on a park bench—so what is to become of our tramps through Melbourne, Victoria? All of this was unthinkable a few months ago. Today the

© The Author(s) 2020
S. Alexander and B. Gleeson, *Urban Awakenings*,
https://doi.org/10.1007/978-981-15-7861-8_15

curtailment of individual liberties is the new normal. Citizens endure home detention, consumers face rationing, workers accept state subsidies. *Homo economicus* is frozen like a bug in amber. Next we'll be queuing for cabbages.

This is the stuff typically reserved for dystopian fiction, not real life, but many other nations around the world are in a similar (or far worse) position to Australia, with more destined to follow as the COVID-19 virus continues its extraordinary disruption. As we write, the date on the computer reads 1 April, usually a time for jokes and pranks. We hesitate for a moment: Is this for real? Surely someone is playing us for fools. But this is no joke. We are at what seems to be the beginning of a turbulent period whose duration is impossible to forecast. We certainly won't pretend to fully understand what is happening, and none of us can foresee how this crisis will unfold and what changes it will bring to the world, including its political economy, which for decades has been firmly framed by the diktat of neoliberal globalisation.

For all we know we are writing from within a relative calm that could yet prove to be the eye of an even more transformative hurricane. What if the virus mutates and comes back with a vengeance? What yet lies in store for the so-called developing world? Will the 'old world' (the pre-crisis order already seems strangely distant) simply 'snap back', as willed by one of its chief spruikers, Prime Minister Scott Morrison? It is a time of promise and potential, but also great risk. Apparently, workers and businesses will have to accept the cessation of state support in six months, and the clock's already ticking. Good political luck with that! The point is, though, that this is already looking like a time of radical opening in thought and action. Even conservatives are shifting ground and rhetoric. For now, all we can do is nod approvingly at the words degrowth scholar Jason Hickel recently cast out into the Twitterverse: 'Capitalist realism is over. Everything is thinkable'. Indeed it is.

An urban tramp was planned for this coming Sunday afternoon. We were going to apply a chaos theory approach and wander Melbourne's inner city in whatever direction the 'green man' at the intersections ushered us. We would keep walking in that direction until we were stopped at another intersection, and then be guided by the algorithms of the green man again. The idea was to have our urban excursion governed

15 Virtually Tramping Through Post-Normal Times

by some arbitrary external variable, rather than be designed by ourselves, just to see where we would end up. Perhaps somewhere our conscious minds would never have led our feet.

Then again, perhaps we humans aren't as free as we think we are, oblivious to the subconscious drivers that shape our behaviour even without playing games with the 'green man'. On the other hand, as philosopher Michel Foucault would have it, perhaps we humans are actually *freer* than we think we are, blind to the ways our ideas and habits of mind are our most oppressive prisons, suggesting liberation awaits those who are able to shake themselves awake from their own comfortable assumptions. Or, as the case may be, rather than shaking ourselves awake, perhaps we will be shaken awake, violently, by this pandemic which is in the process of casting the world and our place in it in a new light.

Despite our plans for an urban tramp, over the last couple of weeks the social and political response to COVID-19 really ramped up in Australia. Vladimir Lenin is said to have remarked once that 'there are decades when nothing happens, and weeks when decades happen'. We seem to be living through such weeks right now. Our government has proclaimed that all discretionary gatherings be cancelled, people should not travel overseas, and generally everyone should practise 'social distancing'. What a disturbing phrase! For decades social researchers have been publishing studies on the decline of real-life human connectivity and community in Western societies, only for this dastardly virus to emerge out of nowhere and provoke deliberate isolation for the common good. We wonder what the Minister for Loneliness in the UK has to say about this.

What fools we are to try to anticipate the future. Let's just admit that black swans probably lie around every bend in every river; perhaps even around every corner of every suburb. Just when we think we've got everything figured out—one day, if we ever manage to finally transition to clean energy and share the world's resources equitably—no doubt an asteroid will hit, or the Singularity of artificial intelligence will arrive, or we will discover that Trump has amassed a clone army. What we can be sure of is that the rock will fall down the other side of the mountain and our Sisyphean labours will begin again. *C'est la vie.*

Anyway, we cancel our tramp, just to be safe, pending further thought and discussion. But what to do instead with the free afternoon? What would Thoreau have done? Naturally, he'd have gone for a solitary walk in Walden Woods, having socially distanced himself by choice, living on the shores of the pond 'a mile from any neighbour'.[1] But we urbanites don't have that romantic option, nice as it sounds, and with the virus spreading like lightning, the idea of hopping into the confined spaces of a train or bus to exit the urban landscape isn't too appealing. Today, at least, we decide to stay at home. Frustrating though this is, a tweet quickly gives us some perspective: 'Your grandparents were called to war. You are being asked to sit on your couch. You can do this. #quarantinelife'. For a time we watch history unfolding online, forgetting that we are not observing it but participating in it.

A perverse idea comes to mind: In this technophilic age, what are the prospects of going for a tramp online? Surely some virtual reality guru has already developed some such programme or app. And then, something re-emerges from the depths of fading memory: a couple of years back a student of ours mentioned that the story of Henry Thoreau's life in the woods had been turned into a computer game, aptly called, *Walden: A Game*.

Anyone who has any appreciation of Thoreau's love of the wild outdoors would be utterly appalled and outraged that someone would even *think* of making this game; a game in which an avatar of Thoreau walks 'virtually' through a pixelated Walden Woods to 'experience' something of his life. It is so antithetical to his biophilic sensibility that the game's concept is almost offensive; an act of 'quiet desperation' as Thoreau would say.

Still, as one online reviewer of the game noted, 'It is deeply ironic that people who will scoff and criticize those who would play this game instead of going outside would *never* criticize those who stay indoors to read a book'. Fair point—raising enough doubt to give us the courage to try it out. With a Sunday afternoon free, intrigued at the prospect, we download the game—solely in the interests of critical scientific inquiry, of course. Nevertheless, as the game loads, we sense our friend Henry turning in his grave. Apologies, Comrade, we are but urban researchers in search of enchantment, and today a disturbing global pandemic has kept us indoors. Forgive us our sins.

15 Virtually Tramping Through Post-Normal Times   163

Courtesy of Tracy Fullerton and the Walden Team ©

It turns out the game isn't too bad, although surely it is the slowest game ever invented—no doubt by design, a lesson in mindfulness. There wasn't a single car chase or explosion. Instead, as Thoreau's avatar, you finish building your cabin and plant some beans; you quietly fish in the pond and have conversations with Ralph Waldo Emerson. Quotes from *Walden* pop up here and there to give inspiration as you saunter through the woods and appreciate the wildlife. Enjoyable enough for a short while, and a valued distraction from virus-talk, but surprise, surprise, it ain't no actual walk in the woods. Your tramps tire of the game and before long we unplug our consciousness from the computer.

We leave that alternative reality, however, only to discover we are still in an alternative reality. What to do now in these times of social distancing? In true postmodern fashion we turn the computer back on and check the news. The real world returns at once. In *Walden* (the book and the game), Thoreau had spoken of the necessities of life—shelter, food, clothing, and fuel. It seems what he forgot to add is toilet paper, as certain members of society would insist in this time of COVID-19 hysteria. Undoubtedly the most disturbing aspect of COVID-19 is not the virus itself but the ways in which some people have responded to it. Surely hyper-individualist 'panic buying' is not the best way to proceed, although this speaks to the insecurity induced by a neoliberal age where social safety nets have been hollowed out over recent decades. At the time of writing this entry, our supermarkets literally have no bog roll. We also read how, in the USA, gun sales are skyrocketing.

We note further the disturbing stories of people hoarding trolleys full of hand sanitiser. While recognising the natural and understandable insecurity induced by the coronavirus, such accumulation seems particularly nonsensical. Presumably the people buying all the sanitiser are motivated by the desire to avoid the virus. Fair enough. But if no one else can buy a bottle of sanitiser while a few people have accumulated it all, then the virus spreads more quickly than if we had a bottle each. We see this as an apt analogy for a broader lesson in how distributive equity can be better for us all. In fact, if you're upset that some people took all the sanitiser, wait till you discover that a small handful of men have taken half the world's wealth.

Is that where neoliberalism has taken us? Towards practices of extreme insecurity and selfishness, where we take everything we can for ourselves, the rest of society be damned? When responding to crisis, so much depends on the state of our minds and the strength of the social fabric. We can descend into barbarism or come together in solidarity.

In dealing with this COVID-19 crisis and future crises, let us take social inspiration from the Italians. When one street isolated themselves in their homes to minimise contagion, in a now-celebrated story they sang together one evening from their windows. If we start in that spirit, we can face anything. If we start by hoarding loo paper and guns at the slightest bump in the road, we may well end up in a dystopian future resembling Cormac McCarthy's *The Road*. There are various ways to respond to crisis, and the same crisis can be experienced very differently depending on the values that shape the social response. The choice is ours, if we choose it.

The burdens of freedom can be exhilarating, while also weighing heavily on human shoulders. That weight is always preferable, of course, to the chains from which it arose.

Note

1. Henry Thoreau, *Walden*, in Carl Bode (ed.), 1982. *The Portable Thoreau*, p 258.

16

A Time for Bad Poetry

As we approach May Day in Melbourne, the Great Disruption of COVID-19 continues to widen and deepen social injury to an extent not seen in these parts (or many others) since the Great Depression. On this day (4 May), the Melbourne *Age* newspaper reports the cries of social advocates who fear for the most vulnerable in this outbreak of disease and penury. Many have been left outside the boundaries of newly expanded public relief—disabled people, international students (here at our invitation in vast numbers), asylum seekers, and the hosts of others who simply fail the rigid new tests of eligibility.

Australian Treasurer Josh Frydenberg shrugs such considerations away, observing the government 'had to draw the line somewhere'. Yes, lines are being redrawn and the tramps are not alone in thinking that a great new round of injustice is about to be served down on the vulnerable. The worst off are being made even worse off. We think of the Bob Dylan song line about down and outers, 'When you think you've lost everything, you find out you can lose a little more'. All of this is preventable if the rich were required to carry a fair social burden. In these neoliberal times, however, their vast wealth hoards are to be strictly protected from the claim of redistribution. The poor it seems will be pushed down further by the crisis into the deepening twilight of injustice. Dark, not just diseased times seem upon us.

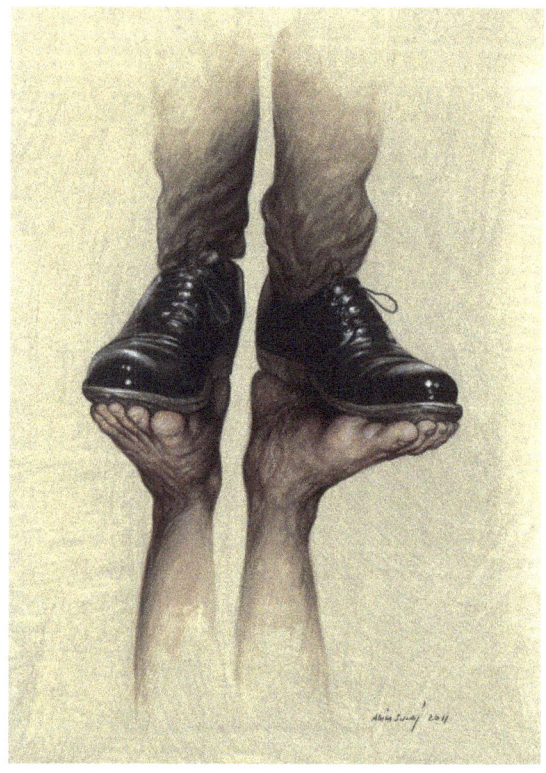

Courtesy of Agim Sulaj © (http://www.agimsulaj.com/).

In his well-known verse sequence 'The Darkest Times 1938–1941', Bertolt Brecht declared that a 'Bad Time for Poetry' had arrived in the world.[1] The simultaneous ascendancy of violent totalitarianisms in Germany and the USSR and, consequently, the opening of wider global war, were the main cascading stages of the dark times, which had emerged snapping at the departing heels of another terrible period, The Great Depression (1929–1933). All this had knocked the poetic stuffing out of much of the world and Brecht gave voice to this great exhalation of literary animus.

Just one complicating fact—Brecht expressed this in verse, thus demonstrating in that dark moment the unvanquished power of poetry to convey great human affect. Not all poetry of course; the high florids of romanticism, for example, were hardly the songs you'd want to hear in a time of peril. Looking back in 1951, Theodor Adorno exhaled that 'there can be no poetry after Auschwitz'. He surely meant that 'good' (i.e. lyric) verse was impossible, not that poetry itself had lost its historical licence.

Writing later in 1976, the astute critic John Willett called Brecht out: '…in a sense he was deceiving both us and himself… for he did go on writing rhymed verse, however incongruous it might seem'.[2] We think there is more to say on this. The Darkest Times indeed witnessed a declaration of violent total war on the human Enlightenment and much of its artistic and literary forms. But Brecht went further than this observation to doing something about it, showing through his own writing the heightened capacity of poetry to light lamps of courage and interpretation in dimming times.

Fast forward to COVID-19 2020 and a dark unfolding cloud that promises to stay with us for years to come. The heretofore busy world has dimmed, not just through human death and wider suffering, but also as economic activity has crashed and movement has stalled, meaning the lights have literally gone out in cities across the globe. Your tramps think there are 'brass linings' to these clouds, and we observe their shimmering forms through busy news reportage and our own repurposed (for exercise) meanderings through Melbourne under lockdown. Traffic has slowed, birdsong has lifted, people are out with their children on what should be work/schooldays, and there is a constantly reported stream of small acts of human connection and solidarity in the midst of the social fear and (potential) loathing that a pandemic must bring. The ecology of a planet bruised by the globalised growth machine is being given a breather, while humans have been released in many cases from the treadmills of neoliberal work regimes (yes, many to the vicissitudes of un(der)employment and financial stress). More on that in the pages to follow.

For we in Australia these new Darkest Times have, as in Brecht's day, unfolded as a series of disasters, of which COVID is the latest stage—first the prolonged and unprecedented drought and hot conditions of recent years, then the Great Conflagration of the 2019/2020 bushfire season watched on by a startled world, followed closely now by the pandemic. The great cloud of smoke generated by the Australian bushfires is still making its way around the globe in the time of pandemic respiratory disease. The term 'biblical' sits on many lips, mindful of the great book's poetic testimony to a human history afflicted by epic and deathly disruptions—plagues, floods, wars, you name it and the book will say it. It even ends with a bodice-ripping final book, a revelation that history will end one day in a great consuming apocalypse. That the human condition is forever overshadowed by the lurking threat of disaster is poetically captured in the much-intoned Old Testament of Job 5:7:

> Yet man is born unto trouble, as the sparks fly upward.

And yet somehow, here we still are, on the ground of Earth, in vastly larger numbers than ever, far from extinguished by all the blows that have rained historically upon us. The Victorian industrial city, and Melbourne is a prime child of this form, witnessed unspeakable ravages from diseases intensified to ghastly effect by the appalling living and working conditions of laissez-faire capitalism. For a time, before proper sewerage and drainage was installed from the 1890s, Sala's 'Marvellous Melbourne' was popularly scorned as 'Smellbourne'—its marvels equalled by its excrescences in the minds of residents and visitors alike.

Our minds are drawn back with this historical thinking to the dead and their many urban graveyard estates, a subject of interest in the before-COVID part of this book. Walter Benjamin, admirer and retailer of Brecht, wrote in the very midst of the Darkest Times (1940) that '…even the dead will not be safe from the enemy if he wins'. The enemy was fascist capitalism; a system that used industrial violence to prosecute accumulation as well as many (potentially countervailing) cultural ambitions, such as the extermination of unwanted peoples. In words often attributed to Lenin: 'Fascism is capitalism in decay'. Later, it was clearer that Stalinism was no less an enemy than fascism to human prospects.

16 A Time for Bad Poetry

ONE MARKET, ONE CONSUMER, ONE LOGIC, ONE DIMENSION

Anonymous ©

We take on Benjamin's stern warning in these times where the enemy of human prospect remains, as it was for him, capitalism, but in new historical drag and with a subtler (less outwardly barbaric) script, neoliberalism. This late, *neo*, form of long failed laissez-faire reminds us that our forbears' struggles for justice are not safely enshrined from incorrigible capitalism which seeks ever to assert the free rein of the market. We accept Benjamin's assertion that the legacy and good name of the dead must be asserted and not assumed. We must keep them safe by keeping them near us… in our hearts and minds. And for the tramps, in our meandering, inquiring gaze.

Thinking this, we decide to revisit the dead, surveyed earlier in the book (see 'Grave Matters'). Under COVID they have already been badly treated (yes, with some great necessity). The hurtful, if necessary, impertinences are many. Bedside deaths of virus sufferers cannot be attended by families, and their funerals are limited in numbers to a handful of grievers. We can only imagine a sea of heartbreak flooding outwards as the ordered (indeed industrialised) protocols and rituals of human departure are swept away by the pandemic. In the time of COVID, we fear for the wellbeing of the dead because they represent everything presently that the social will is opposed to, understandably in a time of pestilence and deadly threat. Swept into the makeshift morgues (thankfully not here) and the shadowlands of lonely home and hospital deaths, the integrity of their testimony to human value and experience, which we strived to uncover earlier in the book, seems under threat by their further erasure from a panic-stricken society.

A wild thought takes us. In a time seemingly bad for poetry, that most intimate form of human affect, could the dead offer up something new? The cemeteries we earlier visited were textual landscapes, vast assemblages of headstone inscriptions bequeathed by the grieving but also sometimes by those prescient departed who planned funerals and wrote epitaphs; altogether a giant volume that surely evokes the human condition and its deathly necessity. Much of the headstone text is poetically expressed in a variety of forms which can be distinguished: staid selections from sacerdotal and classical texts, the folksy reminiscences, the heart-rending *cris de coeur* of parents for lost children, etc. We wish to swiftly survey this sea of text for wisdom in troubling times, attempting thereby to uphold the dignity and the authority of the dead to teach us lessons through the legacy of their funereal poetry.

We are here (again) at The Melbourne Cemetery; at an entrance whose main gates still host banners (observed in this book) offering last 'land sales' to canny customers. Under the volcano of COVID-19, everything we observed is starting to seem doubly ironic. Land deals in graveyards assume a special poignancy in this deathly moment. Enchantment is taking newer deeper forms in an old way, by 'land grabbing' whatever new unexploited morsel of the world can have a sale value put on it. Proceeding up the early thoroughfare of this giant boneyard, a state government banner announces the pandemic's arrival at the field of the dead. Nothing and no one are spared in a time of crisis, including, as Benjamin observed, the dead.

16 A Time for Bad Poetry

We are looking for poetry in dark times. Our healthful tramp through this sprawling, ethnically polyglot resting ground takes us through a Babel of verse; inscriptions in modern and ancient Greek, Latin, Italian, Yiddish, German, and English of varying quality. Sadly, our risible language skills block the way to this rich field of language dreams. Then our procession encounters on one headstone the beautiful single word poem of 'Derrimut', a lone ancient word in this great book of death written in settler languages. Here lies since 1864 the 'Native Chief' (as the inscription reads) who saved the first colonist invaders from a defensive massacre planned by 'up-country' tribes. An unconsciously absurdist inscription on his stone notes that 'Derrimut closed his mortal career in the Benevolent Asylum…aged about 54 years'. Ah, is this condescending whimsy? Or more likely the language of a settler capitalism already assuming that every biography, including indigenous, fitted the industrial script. Absurdist Victorian poetry, our first finding.

Further…we trawl with free association the headstone texts, open to poetic impulse. We pass a plinth marking the death of a woman who died in 1897, aged thirty-seven years (not a public figure so we choose not to share her family name).

> Leaves have their time to fall.
> And flowers to wither at the north winds [sic] breath:
> Each season has clime.
> But who will teach us when to look for thee. O Death.

This stands out to us; a woman is remembered epically, in anticipation perhaps of Brecht's project, where drama is less about script than the lesson, in this case, the deepest human necessity, acceptance…of mortality, finality and even peace—values buried in the whirring rhythms of the industrial city. We move on…moved.

Next encounter, Julian Thomas (1843–1896), by pseudonym the 'Vagabond', slum journalist of Victorian Melbourne who wrote about and aroused much sympathy for the city's outcasts. What deathly poetry is offered here? After a notorious journalistic career, his real identity always hidden under the Vagabond pen name Thomas, son of middle-class English parentage, eventually died in Melbourne penniless, yet his funeral caused an outpouring of public grief and was attended by multitudes. Thomas was remembered as a friend of the poor, his regularly journaled 'slum investigations' for major Melbourne periodicals arousing much sympathy for the cause of social justice (though it didn't have that name then).

Thomas's memorial was raised by friends. Near the base of his grave plinth we read the Shakespearean invocation, 'After life's fitful fever he sleeps well'. In feverous times, it's good to be reminded by Thomas that the sleep of death, once finally achieved, is good and deep. An inevitability, a knock in the heart of the sleepless growth machine of capitalism. The Vagabond, or at least his testifiers, gives us one final note of assurance in deathly times, a side tablet recording 'To live in hearts we leave behind is not to die'. We are not immortal, but we can find solace in the enduring legacy of love, or at least remembered affection.

Another path is taken in this death maze and we come to the remains of Marcus Clarke (1846–1881), English-born colonial litterateur. He wrote many good reads, but we recall him especially for his early invocation of the 'weird melancholy' of the Australian landscape, which had heretofore disturbed without such faithful expression the European eye. We find again on his headstone Shakespeare's gratification for these dark and deathly times. Like Thomas, he sleeps well after a fitful fever. Another fever tag in COVID times. Who can say where these folksy valedictions were drawn from, their provenance is generally not recorded, but should they be taken otherwise, as expressions of sentiment and hope in the after wash of death? We think it a graven language for the times.

Lastly, unexpectedly, the impressive grave of 'James Henry Scullin' appears. Scullin was a Labor Party Prime Minister from 1929 to 1932, a hard watch we venture to say. A progressive political party left to deal with the catastrophic regressions of capitalism. Scullin's time was overshadowed and immensely politically complicated by the breaking crisis of the Depression. He fought hard for the cause of social justice but had to give way through politically pragmatic accommodations to the stubborn, class born(e) refusal of the rich to ever pay for their crimes through economic redistribution.

In stone, Scullin's epitaph (in his own remembered words) asserts:

> Justice and humanity demand
> interference whenever the weak
> are being crushed by the strong.

We're stopped in our tracks. The dead have spoken today, several times, and now in this clear poetic invocation. Justice and humanity demand a place at the pulpit. We fear the injustices of these COVID times, when the urgent claims of human frailty and the necessities of social solidarity are left to bawl against the hard-shut windowpanes of history. The conservative Australian government has firmly committed itself to a 'snap back' of economy and society at the earliest possible point in the crisis. What utterances could break through this defensive farming of neoliberals?

The tramps wonder about what they have heard today, the invocations of the dead. None of what we have surveyed and read today would amount to 'good poetry'. This bricolage of unvarnished and heartfelt human exhalations resounds with social affects—affirmations of human frailty, belonging and dependency, buttressed by aphorisms on 'life's lessons'. It's a bequeathed historical truth worth taking seriously. None of it would have interested poetry prize committees historically, and certainly not now. It is in brute fact a vast concoction of 'bad poetry' written in the boneyards of human fate. It might be what we need in these Dark Times.

Notes

1. Bertolt Brecht, 1976. *Poems 1913–1956*. London: Methuen, pp 321–360.
2. John Willett, 'Disclosure of a Poet' in Brecht, 1976. *Poems*, p xxi.

17

Shimmering Text: Re-Reading *The Plague* in the Coronaverse

Walking a novel path through the city can offer mystery and intrigue, as one absorbs whatever new urban experiences are on offer in all their glorious diversity and unique grit. 'Let's take this side street', the tramps would often say, just because we hadn't sauntered that part of the concrete jungle before. On the other hand, sometimes a familiar route through one's own neighbourhood offers rewards too, like the embrace of a close friend, rich *because* it is familiar.

As an urban Thoreau might have advised, don't bother sauntering to the other side of the city (let alone the other side of world) before you've properly explored your own backyard. Indeed, to paraphrase an instruction from one Thoreau's townspeople William Ellery Channing: before doing anything, make sure you've *devoured yourself alive*. We would only add that there is always inner work to be done—and that is the beginning of politics.

So too with the great works of literature. An excursion through the pages of a worthy new book has well known aesthetic and intellectual delights. What is acknowledged less often is how reading an excellent book *more than once* can be surprisingly enlightening too. Rarely, if ever, does anyone read a book or essay as carefully as it was written, which means it should come as no surprise that a second or third reading can often bear delicious fruit that one missed on the first pass through.

Even the closest reading of a well-crafted text does not exhaust its potential to enchant or disturb. Why? Because either the world or the self (or both) will have evolved in various ways—meaning neither the reader nor the book are really the same entities they once were. They both become different things in a different world, such that, in a sense, they have never really crossed paths before. One always returns to a book with a fresh, more refined perspective. Even though the pages may look identical to the undiscerning eye—the same black print on the same white pages—the meanings of a text shift with the changing times and our changing selves. If you look closely enough, even the text itself sometimes seems to shimmer.

Your tramps learnt these mysterious lessons once again in the midst of the COVID-19 pandemic, as we decided (not without a sense of foreboding masochism) to re-read Albert Camus' *The Plague*.[1] Some people returned to the movies *Contagion* or *Outbreak*; we returned to *The Plague*. Many readers will know that this book tells the story of a city beset by a disease that had transferred from animals to humans, only to turn the city upside down with a flood of unprecedented challenges—including isolation and lockdown, of which we will soon give account. If it is a time for bad poetry, perhaps it is a good time for literature.

We had read the novel ten or fifteen years ago but couldn't resist the perverse temptation to read it again in the midst of our own plague. How have we changed since that first reading? More importantly, how will reading the book *now*, in the Coronaverse, change how the reader digests it? In this chapter we tramp, not through the streets of Melbourne, but between the beautiful, disturbing, and yet edifying sentences of Camus' *The Plague*, that is to say, through the city of Oran. The similarities of the setting and situation, as will become clear, prove to be more remarkable than the differences. A book within a book, we present the analysis in two parts.

Part I

Let us begin 'before plague'. The city of Oran is described by the narrator as merely a large French port on the Algerian coast, a colonial settlement

17 Shimmering Text: Re-Reading *The Plague* in the Coronaverse

distinguished only by its ordinariness. So ordinary, in fact, that everyone agreed that the extraordinary events that took place there seemed, as it were, out of place. It could be any city of industry and trade around the world, increasingly without birds, gardens, or even the rustle of leaves. In short, Oran was an ugly, thoroughly negative place. Disenchantment came naturally within its gates.

We are told that the citizens work hard, 'but solely with the object of getting rich'. Their chief interest is in commerce, and their chief aim in life is, as they call it, 'doing business'. With a modest qualification the narrator admits that the people of Oran 'don't eschew such simpler pleasures as love-making, sea-bathing, and going to the pictures'. But we are told they reserve these pastimes for the weekend and employ the rest of the week 'in making money, as much of it as possible'.

These pursuits, practised with a feverish yet casual air, are not peculiar to the city of Oran, of course. It could be any contemporary industrial settlement. In other words, the city was 'completely modern' and after a while, the narrator warns, 'you go complacently to sleep there'. Sleepers, wake! In the rush for material riches and accumulation, one gets the sense that the forces of capitalist dehumanisation had crept into the town—whether it be Oran or Melbourne or the reader's own city—well before the plague arrived. It is little wonder that the city itself is sometimes described as one of the book's central characters, along with the narrator and the plague itself.

We all know that pestilences have a way of recurring in the world, yet the narrator accurately notes that 'somehow we find it hard to believe in ones that crash down on our heads from a blue sky'. They always take people by surprise, and so too with us. Camus' story, no less surprising than our own, begins as Dr Bernard Rieux, a central figure in the novel, stepped out of his surgery only to discover a dead rat underfoot in the middle of the landing. Without giving it further thought he kicked it to one side, but as he was leaving the building the doctor mentioned the incident to the door-porter and kindly asked that he saw to the rat's removal. 'There weren't no rats here', the door-porter replied, resembling the knee-jerk reporting of the Murdoch-dominated Anglophone media like *The Australian* or *Fox News*. These early signs of plague must be some sort of democratic or liberal hoax, like climate change. There ain't no such thing.

In vain Dr Rieux assured the door-porter that there *was* a rat, presumably dead, but the man's conviction wasn't to be shaken. 'There weren't no rats here', the door-porter repeated, with a closed mind. But if there was a rat, he added, it was brought in from the outside. We are reminded again of the news reporting in recent months with all the crypto-racist undertones about the Chinese origins of COVID-19. Sadly, the reader soon discovers that the door-porter is the first person to die of the plague, despite there being no rats.

What began with one dead rat soon became a disturbing nuisance as numbers multiplied by the day, gutters and dustbins full of them, blood spurting from their mouths. The rats became a great topic of conversation, and 'wild rumours' of the cause and meaning of the phenomenon abounded. People began to die, especially the poor, in torturous ways and in growing numbers. In usual classist fashion, when people who were *not* poor also began to get ill and die, it was then that fear truly set in. Even Tom Hanks could get it! 'A wave of something like panic swept the town', and after the disease was named 'there was a demand for drastic measures' and 'the authorities were accused of slackness'.

As the infection rate and death count both rose in 'geometrical progression', emergency measures were contemplated. The townspeople were advised to 'practice extreme cleanliness' and 'households were ordered to promptly report any fever diagnosed by their doctors and to permit the isolation of sick members of their families in special wards'. Furthermore, all those who had been in contact with patients 'were advised to consult the sanitary inspector and strictly follow advice'. The people of Oran didn't have the technological capacity to develop and issue a surveillance app, as we have done here in Australia, but they did their best to monitor the 'strange malady'—the danger of which still seemed 'fantastically unreal'.

Such efforts, however, were unable to stem the flow of infections and there was no vaccine. Medical supplies became scarce and often insufficient. Before long the hospital wards reached capacity, then the cemeteries, and even the crematoriums were struggling to keep up with the influx of plague-ridden bodies. The formalities at funerals were whittled down and of necessity conducted at lightning speeds, often without families present, to minimise risk. At first, this was a cause of social outrage, but in time, as more pressing needs for survival emerged, 'people had no time to think of the manner in which others were dying around them'.

Dr Rieux, who was a prime adviser to the government on medical questions, became conscious that 'the slightly dazed feeling which came over him when he thought about the plague was growing more pronounced'. Finally, he realised what he meant: 'simply that he was afraid'. The doctor worked extremely long hours combating the disease, distancing himself emotionally from the pain and suffering of its victims so he could continue his work. 'There's not a question of heroism in all this', he tells his friend. 'It's a matter of common decency'. That's an idea which he thinks might make some people smile or scoff, but it's the only thing he believes can combat the plague: common decency; helping each other out; doing what one can, even or especially if one is facing a 'never-ending defeat'.

Occasionally there were days when only a few deaths were reported and people began to wonder whether perhaps the spread of infection was beginning to wane. But then the death count shot up vertically. Finally the authorities got alarmed and jolted into action. An official telegram read: '*Proclaim a state of plague Stop close the town*'. Suddenly citizens woke to discover that the city was in lockdown. The borders were closed and 'commercial activity ceased abruptly' and 'no vehicle had entered since the gates closed'. Traffic thinned out progressively until few vehicles were on the roads (or in the air); most shops were closed, and others began to put up '*Sold out*' notices while crowds of buyers stood waiting at their door. Naturally, the epidemic 'spelt the ruin of the tourist trade', and more generally everyone testified that commerce itself 'had died of the plague'. We trust this all sounds rather too familiar, and the similarities do not end there.

The authorities soon became anxious about food supply, and profiteers were offering, at enormous prices, various essential foodstuffs and products not found in the shops. The result was that 'poor families were in great straits, while the rich were short of practically nothing'. That said, 'the plague was no respecter of persons and under its despotic rule everyone, from the Governor [think Boris Johnson] down to the humblest delinquent, was under sentence and, perhaps for the first time, impartial justice reigned in the prison'. There was a story of a grocer who had laid by masses of tinned provisions 'with the idea of selling them later on at a big profit'. When the ambulance arrived to take him to the morgue, several dozen tins of meat were found under his bed. A rather unpoetic justice, one might say.

In a similar vein, peppermint lozenges had vanished from the chemists' shops, 'because there was a popular belief that when sucking them you were proof against contagion'. Better to consume mints than bleach or disinfectant, we suppose. Or better still, as the citizens of Oran (and Australia) would try to show: 'The best protection against infection is a good bottle of wine [or whisky], which confirmed an already prevalent opinion that alcohol is safeguard against infectious disease'.

On that last point, your tramps remain without medical expertise but cautiously optimistic, and it must be confessed that we've upped our consumption a touch during the pandemic, just to be safe. So far so good, although we dare draw no inferences and make no predictions. We are merely passionate amateurs when it comes to whisky, but as for the health benefits, it's too early to say. Testing continues.

Part II

Returning to a more serious mood. One of the principal themes of *The Plague*—both Camus' story and its contemporary retelling in the Coronaverse—is the torment of separation. Dr. Rieux's wife had left the city for (unrelated) health reasons only days before the city gates were shut, keeping them apart during the epidemic. Similarly, a journalist called Raymond Rambert had been visiting Oran to gather information for a story just as the city went into quarantine. He was now desperate to find a way out to reunite with his love, although his repeated pleas to the authorities proved ineffective. Such separations might feel like a 'special case', he was told, but they rarely are, and the authorities politely but firmly advised that no exceptions were to be made. Rambert, for the time being, was stuck in Oran's sick bubble.

Even your writing tramps share these words (in May 2020) still under lockdown, with state and national borders closed, and physical distancing still a mandatory regulation of ordinary existence. Like the citizens of Oran, we are all 'prisoners of the plague' and in exile, but an 'exile in one's own home'. Essentially, no trains or planes are passing through or overhead. A dinner with friends or a drink at the pub are but figments of the 'before-COVID' imaginary. Conversations at the workplace, however fleeting, are sorely missed. In retrospect, we see so many of life's simplest and richest of pleasures were taken for granted. As one citizen of Oran was to admit: 'We'll all be nuts before long, unless I'm much mistaken'. Mistaken he was not, to which every parent who tried to homeschool during the pandemic can testify. 'Tried' being the operative word. Teachers, nurses, doctors, farmers, and other essential workers: we salute thee.

Burdensome though the disruption was for the citizens of Oran—and for all of us under lockdown today—in this separation people are not alone. Or, if we are alone, we are alone, *together*. The narrator in *The Plague* notes: '…a feeling normally as individual as the ache of separation from those one loves suddenly became a feeling in which all shared alike and – together with fear – the greatest affliction of the long period of exile that lay ahead'. The world may be cruel and repugnant but let us feel and live in solidarity with those who suffer in it. That, fundamentally, was Camus' ethic. His most famous existentialist novel, *The Outsider*,[2] did not uphold such an ethic, and so *The Plague* represents an evolution in the direction of solidarity and participation. Whether it is the plague or COVID-19, a common tragedy makes our collective predicament evident to all, thereby establishing, in the words of one commentator, 'the minimal conditions for bringing humans together in a collective effort'.

We see this in how several characters respond to the ambiguous challenge of the plague. The journalist Rambert, after losing his battle with the authorities to gain permission to leave, is introduced to some smugglers who organise for his escape (at a hefty fee). But on the evening when the escape was supposed to take place, Rambert goes to Dr Rieux and explains that he cannot leave. 'I always thought that I was a stranger in this town and that I had nothing to do with you. But now that I have seen what I have seen I know that I belong in this place whether I like it or not. This business concerns us all'.

The doctor asks about Rambert's need to reunite with his lover and the journalist responds that he'd feel ashamed if he left. Surprisingly, perhaps, Dr Rieux counters, saying that there is nothing wrong with pursuing happiness—at least, he has no arguments against it. 'Certainly', Rambert says, 'but it may be shameful to be happy by oneself'. Here we see one of the most powerful symbolic moments in the book, where a person chooses meaning and struggle over personal happiness, only to discover—promisingly—that a life of meaning and purpose makes one happy, albeit in a different sort of way. One does not have to be a martyr.

Inspiring but challenging in a different way is the character of Jean Tarrou, a good-natured man who arrived in Oran some weeks before the plague and who became a close friend of Dr Rieux. One of Tarrou's central contributions to the story is to organise a grass-roots assistance (resistance) movement to work with Dr Rieux to combat the spread of the infection and help communities in need. We come to understand Tarrou's driving force later in the book when we learn that his father was a prosecuting attorney who tried death penalty cases. After attending one of those trials as a boy, the young Tarrou recognised in himself an innate disgust for the death penalty, which he regarded as state-sponsored murder. Ever since then his moral orientation in the world involved fighting against unnecessary suffering and killing. 'There is something lacking in my mental make-up', he says, 'and its lack prevents me from being a rational murderer'. He even gave up revolutionary activism when he saw the same urge to violence amongst his fellow political agitators. He interprets the plague metaphorically, as much a spiritual disease as a physical one.

'I've drawn up a plan for voluntary groups of helpers', Tarrou tells the doctor. 'Get me empowered to try out my plan, and then let's sidetrack officialdom. In any case, the authorities have their hands full already'. In this engaged spirit, Tarrou forms what come to be called 'sanitary squads', which can be understood as small self-organising activist communities that do what needs to be done to help minimise suffering amidst the crisis. And they prove to be effective, necessary even, especially given an overwhelmed bureaucracy.

Tarrou accuses the authorities of a lack of imagination in their response to the plague. 'Officialdom can never cope with something really catastrophic'. Fiercely moralistic, Tarrou says he is seeking peace by trying to

17 Shimmering Text: Re-Reading *The Plague* in the Coronaverse

become a 'saint without God'. Less ambitious but equally committed is Dr Rieux, who works tirelessly to reduce suffering where he can, just because it is the decent thing to do.

Neither Tarrou nor the doctor subscribe to the Christian response to the health crisis, which involves organising a 'Week of Prayer'. Father Paneloux is a well-respected Jesuit priest, known for giving powerful and chastising lectures. In his first sermon after the plague arrives, he berates his congregation for their laxity and declares that the plague is a just punishment. 'Calamity has come on you, my brethren, and you deserved it'.

Later in the book Father Paneloux delivers a second sermon, but by this time his message has shifted. The priest had been present at the death of a small child who suffered violently before dying, and this suffering of an innocent child casts his faith into question. He no longer sees the plague as punishment. But rather than give up his faith he doubles-down, insisting that, while the child's suffering has no rational explanation, that simply means people need to throw themselves unquestioningly into their faith, despite the apparent contradiction of a loving God allowing the suffering of innocents to occur. Soon after this sermon, Father Paneloux develops a condition that, without quite being the plague, shares many of its symptoms. Consistent with his own conception of faith, the priest refuses medical treatment and dies.

Perhaps the simplest and most inspiring expression of solidarity in the novel comes from the old government clerk, Joseph Grand. When a neighbour, Cottard, attempts to commit suicide, Dr Rieux looks around for someone to keep an eye on him. Without thinking twice, Grand is there to lend a hand. 'I can't really say I know him, but one's got to help a neighbour, hasn't one?' If citizens in crisis begin with that ethic of mutual aid, then the community will survive a plague, no matter how bad it gets. The suicidal Cottard represents the opposite sort of character. Before the plague he was ready to kill himself, but he is somehow uplifted by the onset of the plague and the suffering it causes others. He is happier during the epidemic, because it has made him feel part of the group.

Grand's significance in the novel, however, goes deeper still—beyond his noble assumption of mutual obligation and care. Despite needing to earn his livelihood as a poorly paid government clerk, and being of old age, he nevertheless eagerly commits with 'quiet courage' to assist in the fight against the plague at every opportunity. Given work commitments, he tells Tarrou he can assist from six till eight every evening. Beyond his activism, this humble bureaucrat spends any spare time and energy he has working on a novel. As it turns out, he is also a literary perfectionist (but one without much ability), and for a long time he has merely been writing and endlessly re-writing the first sentence, unwilling to move on until he is sure he has sufficiently polished the opening line. It is absurd, but upliftingly absurd, for it is clear that life, for Grand, is enchanted. He is animated to participate in a troubled world. He does not seek meaning 'out there'. He creates it by living it into existence, here and now, refusing to be dominated by the plague's dehumanising power.

Contrast this with Dr Rieux's old asthma patient, who is convinced that life has no meaning, and so decides the best way to live is to do as little as possible. He chooses to spend his time in bed, mindlessly transferring dried peas from one pot to another, a mechanical activity which happens to tell him when it's time to eat. It is the bare minimum needed to survive. This fellow is also absurd, but decidedly in a less uplifting sort of way. It would be better to call this form of life silly, shallow, or even cowardly. Life in a universe without metaphysical foundations is inherently absurd, according to Camus. What matters then is what we do with our absurd existence.

At a superficial level, *The Plague* is simply an aesthetically satisfying and well-crafted story. But such literal interpretations are rarely what an author intends. Beneath the surface Camus is also pointing to the plague as a symbol for the Nazi occupation of Paris during the war—he began writing the book in the early 1940s. In some regards this is a problematic storytelling metaphor, since by 'naturalising' the Nazis into a disease, the manifestation of fascism became the workings of nature rather than the deliberate choices of humans against humans. It also sidesteps the philosophical problem of whether it is ever justified to use violence to resist evil, which is a central theme in Camus' work.

This problem is highlighted in Dr Rieux's own words: 'What is natural is the microbe. All the rest, health, integrity, purity, if you like, is an act of will'. As we have noted in previous chapters, there are various ways to respond to crisis situations, whether those crises be health, environmental, financial, political, or spiritual. Human societies may not always be able to avoid diseases and natural disasters. It follows that the social and political challenge is to manage those crises with compassion and solidarity when they arrive, knowing that 'no longer [are there] individual destinies, only a collective destiny'. This is the solidarity that flows from separation under lockdown.

In true existentialist fashion, Camus also makes it clear that he is commenting more broadly on the human condition, summed up best perhaps in the words of Tarrou: 'I had plague already, long before I came to this town and encountered it here. Which is tantamount to saying I'm like everybody else'. As the novel concludes (spoiler alert) we learn that Dr Rieux's wife has died from her health condition outside the city walls. Just as tragically, Tarrou is one of the last people to die from the plague, although he doesn't leave the world without a heroic fight. These deaths are Camus' way of saying that, even after beating the plague, suffering is part of the human condition and will remain so.

Nevertheless, what makes Camus' work so edifying and surprisingly uplifting is that one closes the book somehow more cognisant of the simple joys of ordinary existence, like swimming in the sea or having dinner with friends. And one becomes more cognisant of the sublime responsibility of personal freedom, heavy though it weighs on us in an absurd universe. After reading *The Plague*—or living in the Coronaverse—one is less likely to take such things for granted. The simple things, the human connections, are the fabric of life and give it meaning. And they are worth fighting for—together.

Still, in the midst of the pandemic, your tramps can only nod in empathy with Dr Rieux who noted that 'if things go on as they are going, the whole town will be a madhouse'. He felt exhausted, his throat was parched. 'Let's have a drink'.

The tramps abide.

Conclusion

As to be expected, the plague eventually passes and the gates of Oran are reopened. The city bursts into festivities and celebration. So too will COVID-19 one day be a memory and not an immediate or painfully near experience. As always, of course, people will discover that 'destruction is an easier, speedier process than reconstruction'. As Tarrou once opined, the goal and primary ability of governments is to maintain business as usual. But we live in a time where 'bouncing back' to normality is arguably the most dangerous thing of all. This is not to glorify the pandemic; it is simply a warning not to romanticise what used to be. We need as many people as possible enchanted by 'an inkling of something different'. And even though the fight for a new world probably means a 'never-ending defeat', we are animated by Dr Rieux's conviction that this is 'no reason for giving up the struggle'.

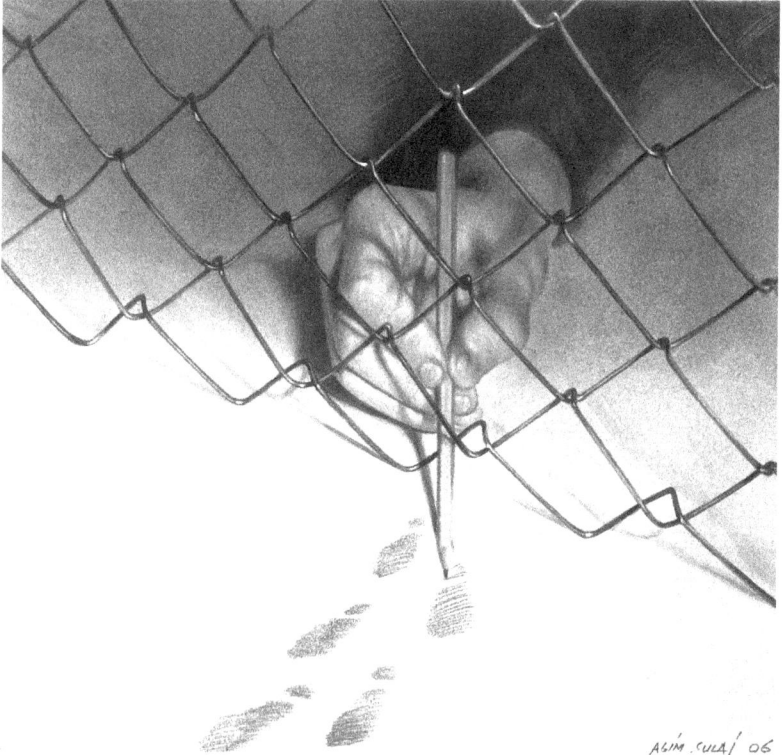

Courtesy of Agim Sulaj © (http://www.agimsulaj.com/).

17 Shimmering Text: Re-Reading *The Plague* in the Coronaverse

Indeed, we write, perhaps, for the same reason Dr Rieux did (who is finally revealed as the narrator): 'So as not to be one of those who are silent, so as to testify in favour of these victims of the plague, so as to leave at least a memory of the injustice and the violence which was done to them and so as to tell simply what one learns amongst scourges, that there are in men more things to be admired than despised'. We reflect on the human response to the coronavirus and feel justified in nodding in agreement. There is more to be admired in humankind than despised.

Brendan Gleeson ©

The Plague ends, however, with a warning: the bacillus of the plague can lie dormant for years 'in furniture and linen-chests' and may again one day awaken its rats and 'send them to die in a happy city'. It follows that overcoming the plague or the coronavirus can never be one of final victory. It merely points to what might need to be done again 'in the never ending fight against terror and its relentless onslaughts... by all who, while unable to be saints but refusing to bow down to pestilences, strive their utmost to be healers'.

For now, however, your tramps—tired of and from the news, and suffering the ache of separation from society—just follow the suggestion of Garcia, who had helped facilitate the plan to smuggle the journalist Rambert from the city:

'How about a stroll?'

Notes

1. Albert Camus, *The Plague*. In Stuart Gilbert (ed., trans.), 1960. *The Collected Work of Albert Camus*. London: Hamish Hamilton. We have decided not to clutter the text hereafter with an endnote for each quote; they are all from this edition and translation.
2. Albert Camus, *The Outsider*. In Stuart Gilbert (ed., trans.), 1960. *The Collected Work of Albert Camus*. London: Hamish Hamilton.

18

Care-Full Times: Suffer the Children

One of the most dramatic and dangerous failings of neoliberalism has been its carelessness, in so many senses of the word. The term connotes recklessness, a defining feature of this political order. Consider how for decades its growth machine economy merrily ran (and still runs) a wrecking ball through the planet's ecology. The Global Financial Crisis (2007–2008), a disorder of its own making, stopped the machine in its tracks for a few years, before it resumed the mindless plunder of natural value. But now, just a few years onwards, nature itself in the form of a virus has called a halt to things again. The end of growth capitalism or merely its suspension? Mostly likely, perhaps, this is just the latest phase in its catabolic deterioration.

As the tramps ponder their work in AC, this carelessness comes to mind. In a sense we have all been handcuffed to a madman (neoliberalism) in recent decades, and now we are uneasily halted with this creature, still shackled but pondering (dare we all?) our chances for escape. Where is the key that we might silently deploy to unhinge us from this unhinged order? This is a big question for all progressives and radicals in this moment. Our guiding philosopher contends that 'enchantment can aid in the project of cultivating a stance of presumptive generosity (i.e. of rendering oneself more open to the surprise of other selves and bodies and more willing and able to enter into productive assemblages with them)'.[1] For the tramps now, the immediate issue seems how to tackle carelessness as a deep social wound, in this contemporary COVID moment. Let's think about that before stepping out again…

The other face of carelessness inherent to neoliberalism is its inability, let alone unwillingness, to value the care of humans as well as ecology. Socially, it is a cruel dispensation that has brutalised the poor, while adding vastly to their numbers, as well as generally effecting a less caring societal attitude towards the vulnerable, including the disabled, the sick, cultural minorities, refugees, and, in many parts, the elderly. Especially in the Anglophone world where it has taken deepest root, neoliberalism has produced a new and lamentable human phenomenon, the precariat, a growing social estate that, while not in absolute poverty, lives nonetheless in a gruelling state of income and employment insecurity. Risk, uncertainty, and vulnerability, as the late Ulrich Beck explained it, are the burning tropes of this age.

Amongst the vulnerable are one estate whose frequent suffering and diminishment has most condemned this reckless machinic order. We speak of children. Incredible to relate, but a system that asserts itself as a natural social form has been eating its children. We do not exaggerate but report the increasingly anxious testimony of health specialists in recent decades who have pointed to the awful fact of 'Modernity's Paradox'.[2] This term emerged in North American health science in the late 1990s and refers to a relentless decline in many developmental indicators for many children in an epoch of unprecedented (if unequally distributed) prosperity.

The alarm sounded by this assessment is coming from medical and human science not social advocacy. Just as climate scientists have been stunned by the indifference of the growth machine to the fact of global heating, child health and development experts have been vexed by its erasure of childhood wellbeing as a social priority. Earlier, poorer civilisations knew that children's development must be a leading concern for any society that wishes a future for itself.

18 Care-Full Times: Suffer the Children

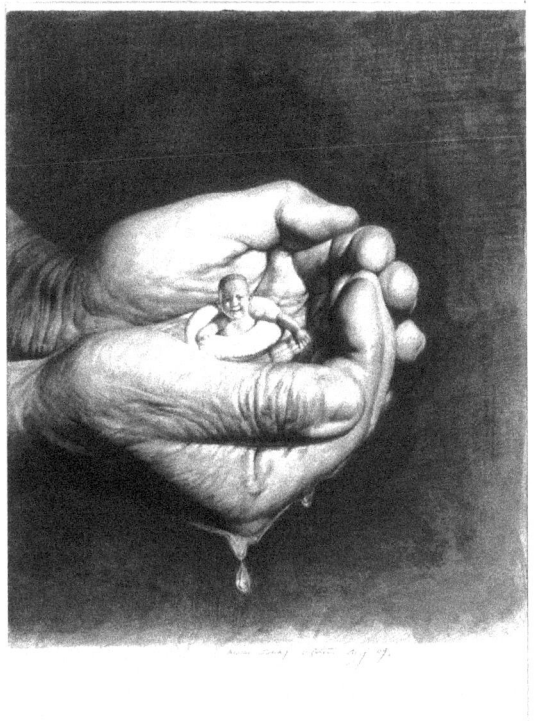

Agim Sulaj © (http://www.agimsulaj.com/).

In their widely read 2005 Australian book, *Children of the Lucky Country?*,[3] child wellbeing experts Fiona Stanley, Sue Richardson, and Margot Prior pointed to neoliberalism's many stressors on children, including poverty, inequality, parental work exhaustion, and, at the other end of the social scale, the lethal downsides of affluence. The book's subtitle showed the scale of their argument and feeling: *How Australian Society Turned its Back on Children and Why Children Matter*. A chilling conclusion of their assessments of Modernity's Paradox was that '[t]he present generation of children may be the first in the history of the world to have a lower life expectancy than their parents'.[4] We trust and applaud their Australian argument more than their wider historiography (some previous societies must have fallen into this extinction trap).

For a time, these brave souls, and their many supporters in the professions caring for children, were able to influence national conversations and gained some attention from policymakers. Stanley, epidemiologist and leading child health advocate, was made Australian of the Year in 2003. The argument was taken up by urbanists and applied to the city, with one of your tramps co-editing the 2006 collection *Creating Child Friendly Cities*, which, as the subtitle explained, was from a group of scholars and experts concerned with *Reinstating Kids in the City*.[5] Many urban dimensions of the paradox problem were essayed, including that the recent growth and enrichment of cities had been accompanied paradoxically by the erasure of children's needs and wellbeing as an urban priority. Urban intensification and densification had literally erased space for children (parks, wild spaces, social facilities, etc.) while rising social and spatial polarisation were condemning many to lives in poverty sinkholes.

In this time, the Australian Research Alliance for Children and Youth was established to progress the cause of childhood advocacy nationally. Considerable interest was generated in professional, industry, and academic circles. Like most such socially progressive debates, however, there was a general blanketing down of discussion imposed by the sudden arrival of the Global Financial Crisis in 2007 and in the general atmosphere of social and economic emergency that it generated. The story from there has been related: neoliberalism climbed down from the gallows of its own making, assisted generously by public bailouts and a failure to sanction the people and ideas most responsible for the crash. The urban growth machines sputtered back into vigorous life as Australian cities were subjected to a new wave of hypertrophic development, best represented in the vertically sprawling towerscapes of the inner cities. These are often poor places for humans, not just children, to reside in.

Fast forward to AC. The COVID virus has not so far generally affected the physical health of Australian children (though globally there have been heartbreaking exceptions to this statistical observation). But along with the rest of society it has overturned their lives suddenly and dramatically, right down to the choreography of everyday routines. How so specifically for kids? The first impacts are those on their parents and carers, many tossed into unemployment, underemployment, or, paradoxically, overemployment for the stressed legions who must now do the work of many. Next on the list of harms, and most obviously, the closure of schools and childcare centres and the general further confinement to home imposed by the great lockdown. Many children have been exposed to the vicissitudes of family stress—homeschooling has generated widespread anxiety and, much worse, there is evidence of rising levels of family violence. A bleak picture indeed.

Or is this the whole story, we tramps wonder? Outside our windows, on the streets around us, in the parks in which we exercise, a different if not completely contradictory reality is suggested. Who cannot see and thus be moved by the hordes of kids ranging about in public space during what are normally school hours, many hand in hand with parents so obviously not at work. Gaze on through time-space to the mornings, early evenings, weekends, when, as never before, families are taking such obvious pleasure in their very existence, going about on walks together, cycling together, or just disporting and gambolling about together on playing fields or indeed any space that will grant freedom to a ball, a kite, or just the act of rolling around. We note with wonder the many fathers and male carers present to the melees, indeed so often central to the horseplay. The blokes are back to the work of care! (We do not diminish the testimony of women's advocates that the great disruption has imposed profoundly unfair and stressful impacts on women.) All this in working-school hours!

With this in mind, we commit to a short COVID-compliant tramp through one small metropolitan patch, the inner suburb of North Melbourne. On its northern edge, sitting in the margins of the glorious Royal Park, is the Royal Children's Hospital, an institution that for one and a half centuries has ministered to kids, and in recent years has surely encountered the meaning of Modernity's Paradox in its clinical and public work. We tramp towards that castle of care, arriving to find it, expectedly, under new strictures of access imposed by the pandemic. Variously positioned sandwich board signs avow 'No Visitors…Parents of Guardians of Patients Excepted'.

We pass a profusely weeping woman comforted by another woman—we guess at a mother and her sister, if looks alike can tell. It reminds us of the vast unfathomable sea of suffering that is human frailty, so awfully present in the sickness and death of children. Much has been rightfully said internationally about the sadness of aged COVID victims dying under clinical isolation without the comforting attendance of relatives and friends. Such sorrow extends to the kids in this vast great castle, with ramparts partially drawn, who cannot be visited freely by many that love them. The feeling is compounded as we pass one of the hospital's playgrounds, now closed with assertive signage, as with all playgrounds. *Achtung, you cannot play!* The tramps pass on wishing all sick kids well, and acknowledging that the crisis has brought new salience, resources, and we hope morale, to the public health sector that cares for them. That must be a good thing…for the cared for.

Brendan Gleeson ©

We swing around through the backstreets of North Melbourne coming up to its popular and normally 'busy bee' primary school. Even the local childfree residents know it. In Australia with our wonderful compulsory voting provisions, local schools like this are scenes on election day of long lines of the citizenry lining up for voting. Everyone gets time to look at the innards of the local school; no bad thing. This school in every sense exudes a sense of warm curation by its community—and who cannot notice the thriving food gardens on its western edge? The visitor of any stripe encounters a flourishing child-centred place.

The school dates from 1874 and has grown in time around its handsome Victorian buildings to be a heavily subscribed and valued public institution in this part of Melbourne. As we healthfully saunter past, it just happens to be 'pick-up' time. Before-COVID, at this time of day, we would encounter the 'ordinary chaos' of the pick-up; cars, kids, adults, crossing supervisors, departing staff, passers through (like us) all converging in an amazing, and daily re-enacted time-space ritual (melee!) focused on safely removing *the child(ren)* to home care. Under the State of Victoria's educational lockdown, the school is essentially closed, replaced, as a fence sign tells us, by 'Remote Learning', even if that general spatial term means here in the dense inner-city schooling from 100 metres away. We walk past the school, a social hotspot eerily deanimated and silenced by the crisis.

Onwards to 'Gardiner Reserve' with its recently and magically renovated playground. Another local kidspace closed by municipal fiat. At the playground, honking public order signs announce the proscription on play, while chequered red tape adds to the sense of a (potential) crime scene. Fair enough, these are serious times. But then we are moved by the sight of one little tot ignoring the dramatic bunting and happily playing alone on a climbing thingy. In the near distance, a sitting parent watches on smiling. Winks all around. The necessary social order is never harmed by such a small, spontaneous instance of refusal. It's how the whole creaking crisis thing should work; a general social ordinance relaxed in the last local instance by safe little plays. Safe safety valve stuff; you know what we mean.

Coming back to the beginning… we realise that the school closure and home lockdown facts still need our concluding attention. We humbly do not have all the answers to the questions this crisis has raised for kids. But see perhaps the runes of answer in its still shifting sands. Our minds are taken back now to a television episode of 'Foreign Correspondent', a regular news magazine shows on the national public broadcaster, the ABC. A few months ago there was a segment that investigated the early and terrible stages of the Italian crisis and its government's hefty if belated lockdown response. Two little five- or six-year-old girls were shown in a video chat. One asked the other something like, '[there are] lots of yucky things… but what do you like about the shutdown?' Her correspondent immediately and enthusiastically offered: 'Ah, we don't have to rush breakfast anymore'.

Jesus enjoined, 'Suffer the little children'. This is from King James' rendering of the Bible (always a story in itself). Suffer in such older English means to tolerate, to be patient, to listen to carefully, finally… just to sit and be quiet with the other. Not social values applied much in recent times to children—especially if we consider the condescending, faux celebration political society bestows on the School Strikes for Climate (Fridays for the Future) only to then promptly get back to business as usual.

Jesus, childless being, said we must suffer children. Strangely (or not) we tramps, apostates but not enemies of the Word, can commit to His view, especially, even fervidly, now in these fractured times. For, in this crisis, what we have (re)learned from his testimony is the *secular* Word of our species, that children must be cherished if you have any wish to the election of history, to the survival of your society. As Bennett aptly notes, 'If all goes well, children seem to be born with a capacity for enchantment, and most adults retain something of this power'.[6] It follows that we have much to learn from children. But we tramps say against softer theologies that other people, care-less people, should not be suffered, even if we dare not call for their eternal damnation.

During the crisis, we hear this civil Word intoned against the reckless cant of neoliberalism. This deathly machine that eats its children, offending earth as much as heaven. The political economist Boris Frankel, in a book title, once observed about the neoliberal ideologues who helped bring this (dis)order into our lives, *From the Prophets Deserts*

Come. 'Curse them back to their deserts!' we say. Let's green our lands, not with profits or prophets, but care.

Notes

1. Jane Bennett, 2001. *The Enchantment of Modern Life: Attachments, Crossings, and Ethics.* Princeton: Princeton University Press, p 131.
2. Daniel Keating and Clyde Hertzman (eds.), 1999. *Developmental Health and the Wealth of Nations: Social, Biological, and Educational Dynamics.* New York: The Guilford Press.
3. Fiona Stanley, Sue Richardson, and Margot Prior, 2005. *Children of the Lucky Country: How Australian Society Turned Its Back on Children and Why Children Matter.* Sydney: Pan Macmillan.
4. Ibid., p 52.
5. Brendan Gleeson and Neil Sipe (eds.), 2006. *Creating Child Friendly Cities: Reinstating Kids in the City.* London: Routledge.
6. Bennett, *Enchantment*, p 168.

19

Rewilding the Suburbs: CERES as a Site of Enchantment

This late afternoon we tramp to and through a place called CERES, a 10-acre 'environmental park' situated in Melbourne's inner northern suburb of East Brunswick, located on Wurundjeri land that runs along the Merri Creek. This is, we feel, one of the most enchanting places in the city, for reasons we will attempt to convey. The acronym stands for the Centre for Education and Research in Environmental Strategies, but it bows intentionally to ancient Greek mythology, to the goddess Demeter (who was known in Rome as Ceres).

Without attempting a detailed engagement with her mythology, suffice to say that this great goddess was a deity of renewal—of birth, death, and regeneration.[1] At times the goddess would descend into the underworld and encounter the shadow side of existence in the realm of death, but through this mortifying engagement she would find new life, energy, and nourishment, returning to the world above restored and replenished. There is a mythopoetic reference here to the changing of the seasons; to the onset of winter as the goddess descends, to be followed by the blossoming of spring on her return. Perhaps it is fitting that we visit CERES on the first day of winter, even if we find her quieter than ever, almost a ghost town, owing to the COVID lockdown.

© The Author(s) 2020
S. Alexander and B. Gleeson, *Urban Awakenings*,
https://doi.org/10.1007/978-981-15-7861-8_19

Your tramps are not goddesses—needless to say!—but we have perambulated through the cemeteries of Melbourne not out of any morbid fascination with death but, like Ceres, to seek light, life, and poetry in dark and bad-poetic times. We have already paid homage to Camus' insight that 'there is no sun without shadow, and it is essential to know the night'. For this reason, amongst others, we find ourselves at home at CERES, as if somehow connected with her mythopoetic roots, deep and distant though they are. It helps also that the people and place are warm and inviting even on the coldest of winter days, like a good mulled wine.

An entire book could be written on CERES, tracing a journey from its conception to its birth and beyond; exploring its achievements and its struggles; its politics and its place. No such comprehensive statement can be attempted here. For now we can only sketch the briefest outline of this complex and compelling project, which was born in 1982, attempting just enough to provide a foundation upon which to offer an analysis of urban enchantment.

Once a desolate wasteland, the site of CERES is now a bubbling hub of environmental and community activism. Its website even testifies to a mission consonant with the present book: 'Our vision is for people to fall in love with the Earth again'. The subtext of this mission is concern over the narrative of disenchantment: we cannot care for what we do not love. So we must learn how to love again, and therein lies the ethical potential of urban enchantment. If we truly love something—if we see in something the capacity for wonder, excitement, and beauty—we'll do anything to protect it. If we feel detached and disenchanted, we'll be less concerned; we'll resign ourselves to the onslaught of industrial urbanism.

The spectacle of CERES is partly due to its urban context. Most readers will have been on a rural farm before; many would have seen a rural ecovillage, either in real life or in pictures. But in this inner suburb, it is inspiring to see the expansive and productive urban gardens, orchards, and chook yards in the midst of an industrial city.[2] As you wander its interwoven tapestry of paths and laneways you discover around every corner a unique mud hut, a quirky meeting place or performance space, or a classroom under the skies, formed around a fire pit. Places seemingly designed for ritual and community abound. Solar panels are perched on every building, many with lichen around the edges

19 Rewilding the Suburbs: CERES as a Site of Enchantment

Courtesy of Laurie Duckham ©

testifying to CERES' long history on the post-carbon transition. Out the front there is even a solar-powered charging station where you can recharge your electric bike or car.

Amongst the range of alternative technologies in play at CERES is a biogas digester, and you'll also pass parabolic solar dishes for concentrating the sun's energy. There is an organic café, market garden, and nursery, which help keep the place self-sustaining, supported by a fascinating array of community education workshops or conversations, anything from permaculture or gardening courses to lessons on how to grow mushrooms or keep bees. Bikes and bees are everywhere.

CERES expands the imagination regarding what an urban landscape and urban community can look like. And it raises the question: What if CERES became the rule rather than the exception? What if these types of projects came to inform a post-development vision of grass-roots urban planning and practice, where communities set out to rewild the city and its suburbs in the same spirit of renewal and regeneration that

Courtesy of Laurie Duckham ©

has inspired the CERES project? These questions seem particularly penetrating in this COVID moment, where everything is thinkable. On this crisp and clear winter morning, we feel, as we do so often, that we are walking through a place that is well ahead of its time, a grounded utopia on the frontiers of deep green ecological practice, one that is calling for the rest of the city to catch up. As Thoreau would say: 'If you have built castles in the air, your work need not be lost; that is where they should be. Now put the foundations under them'.[3]

CERES is a place the tramps have been captivated by over the years. We want to describe these experiences for you, yet in the attempt to do so we find ourselves almost lost for words, as if the experiences we want to describe are somehow touching on the ineffable. Ordinary language somehow seems inadequate, demanding a pure poetic expression. But the poetic is always so hard to craft (especially for academics!), and even when done well it never quite seems adequate to the task it sets itself—of describing the indescribable. In an earlier chapter, we quoted the

19 Rewilding the Suburbs: CERES as a Site of Enchantment

philosopher Wittgenstein, who captures the human condition's linguistic problem in the mysterious phrase that concludes his text *Tractatus Logico-Philosophicus*, often translated as: 'What we cannot speak about we must pass over in silence'.[4]

But our book is *about* those things we cannot speak about... suggesting our project of enchantment was very poorly conceived! We have discovered that when the doors of urban perception are open, sometimes the wonder of minor experiences in the city can teach things that can only be *experienced*, not adequately *explained*. Nevertheless, in this undertaking we set ourselves the meta-challenge of *trying* to talk about what defies being talked about, and thus we quickly realised that, as writers, words were all we had. Without regret, we admit that this book is testament to that flawed and failed (but necessary) endeavour. Some features of the human condition are too important to pass over in silence.

Courtesy of Laurie Duckham ©

In our research, we did fall upon an essay by Australian scholar Freya Matthews, who managed to craft a poetic expression of her own experiences at CERES that we humbly quote at length below due to the poetry of her words. She captures the enchanting capacity of CERES in ways that could flow through our own veins, if not from our less poetic pens:

> Sometimes, at sunset, I meet at CERES with friends, other members of our little group of 'reinhabitants'; we make a fire in the big domed oven to warm the food we have brought to share. We sit in the firelight till late in the evening. Sometimes till later. The moon rises. Windmills swish and trees murmur. There is the occasional clamour of roused geese up at the stables. Close by, but out of sight, the creek glides past in the dark. Bill, our inventor, boils a kettle on a methane gas-ring fed by a dozen barrels of compost. Louis, our musician, strums guitar and sings green-blues. Little rituals sometimes erupt in our midst, to mark the transit of the seasons. We tell dreams, make wishes. The whole site, deserted, shadowy and lamplit, rises and falls around us, breathing, its presence real and palpable.[5]

Matthews begins here with material 'real world' descriptions of people and place, but then moves to a higher discursive plane as she tries (with surprising success) to understand and share the humble yet celestial experience:

> Though the gardens and African huts and animals in their straw beds are all wrapped in sleep, the world itself is awake, alive, alert to the conversation in our circle. And sometimes, sitting there in the company of my friends and this wide-awake world, in this slumbering place, I have a sense of the uncanny, as if the scene around me belongs to another world, a possible world, perhaps the future of this world, but not the present. We are at the edge of reality, neither in the country, for traffic drones in the background, and overhead the sky is lurid with city light, nor in the city, for we are gathered around a camp fire amidst food gardens and paddocks and bee hives.[6]

Her expression then moves from the poetic back to the tone of her other training and capacity as scholar (admittedly, though, this is a poet-scholar distinction she scandalously blurs by somehow retaining the poetic mood):

> Our scene does not belong to the 'developed' world, since we are, at that moment, at the heart of a village surrounded by technologies of subsistence; but nor does it belong to the 'developing' world, since our talk betrays our identities as privileged, white 'first worlders'. This is not a glimpse into the premodern past: there is too much cryptic evidence of contemporary urban civilization – electric lighting, a computer screen glowing through the office window, the power lines. But it is not the face of modernity either, given its animistic ambience. It is a scene cut adrift from reality, and I wonder, where does it belong? In the future? Is this how social life will be organized late in the 21st century? Will we by then have brought nature – habitat and food production – back into the city; will we have recaptured the enchantment of the premodern world, saturated as it was with spirituality, in the secular civilization created by modernity; will we have worked the village culture and human-nature partnerships of the 'third world' back into the alienated fabric of the 'first world'? Or will society in the late 21st century have gone further, much further, down the other path, the path leading to a scenario in which this re-awakened site will be paved over again, sealed up and silenced under storeys of concrete, and the trees and kingfishers, the windmills and honey lane gardens have vanished forever from our cities, the cities which, by then, will have claimed the world?[7]

We are at these crossroads and are in the process of choosing our fate, but late though the hour is, it is not yet too late to change direction. This COVID moment, as we and others have argued, is opening up space for new and hitherto unthinkable possibilities. Which path industrial society takes will depend, we argue, on its cultural mood, the affective state of its citizenry and the nature of its affective propulsions and ethical energy. We thank Freya Matthews for her words, which themselves are enchanting and threaten to evoke a sense of the uncanny. If we could improve upon her expression, we would have done so. Our own experiences of CERES are unique, of course, but Matthews captures 'the vibe' of the place as precisely as language may be capable of doing.

Courtesy of Laurie Duckham ©

As we sojourn through the many curiosities of CERES we are struck by its humble yet enchanting aesthetic. There is a tendency in glossy magazines to portray an image of sustainability that has as much glamour and chic as the luxurious 'high end' of consumer culture, just done in some superficially 'green' sort of way; a hyper-modernist home, governed by that expensive minimalist taste, with a couple of unaffordable electric cars outside. But CERES is not particularly glamorous at all, and yet it is partly that which gives this site of enchantment its unmistakable authenticity. Might this be closer to what real sustainability looks like? A humble and organic connection with people and place?

The community here is clearly participating in the 'salvage economy', making most of their buildings and garden beds out of what they could find, gather, recycle, or repurpose, rather than always buying resources new to make the space look slick and aesthetically homogeneous. Nobody seems that fussed about looking 'brand new', instead choosing an alternative aesthetic of old, second-hand, or homemade clothes, but not for that reason lacking in style.

19 Rewilding the Suburbs: CERES as a Site of Enchantment

All this can make the place look a bit scruffy and patch-worked, but the imaginative work demanded by using reclaimed materials means that each nook and cranny at CERES has its own, unique charm and appeal. Scarcity begets creativity. Whether it is making mud huts from the earth beneath us, erecting garden beds or fences from salvaged wood, or creating shingles for a shed roof out of old tins of oil, there is a down-to-earth but inspired beauty to CERES—an aesthetics of sufficiency—that stirs the soul.[8] This is a community creating the taste by which it will be judged.

The challenge, as always, is to see projects like this not as anomalies in an overwhelming wasteland, but as seeds of the new within the shell of the old. Of course, CERES is not on the precipice of transforming the world, but it is a lived, real-world example that you can touch and see and feel, and therein lies its transformative potential. It shows that 'life can no longer be contained by our conventional scripts'.[9] It offers or prefigures a concrete alternative to the concrete jungle, and walking through a real-world example can provide energy to activists and changemakers. It can make you believe that another world really is possible, and all of a sudden the mood of 'capitalist realism' begins to crumble—the idea that it is easier to imagine the end of the world than the end of capitalism begins to fade. And then the next time you read Ted Trainer's mind-expanding work on the 'Simpler Way',[10] or Helena Norberg-Hodge's incisive and compelling work on localisation, economics of happiness, and 'ancient futures',[11] or David Holmgren's pathbreaking case for a permaculture 'retrofit of the suburbs',[12] you'll have an empowering sense that you've seen something of these visions before, in a microcosm, at places like CERES.

Indeed, as Freya Mathews writes: 'These traffic-free, nature-friendly routes through the city could come to constitute something like a contemporary, urban version of the Aboriginal songlines, sacred pathways for journeying, rather than merely traversing space, and journeying for the purpose of "singing up" the land'.[13] The lesson here is to seek out those sites of enchantment in your own locality; to connect with people involved in practical projects of regeneration and renewal; and together participate in the flow of urban revolt. When energised by enchantment, we come to learn, as 'transition towns' founder Rob Hopkins says, 'that by unleashing the collective genius of those around us to creatively and proactively design our energy descent, we can build ways of living that are more connected, more enriching and that recognise the biological limits of our planet'.[14]

Courtesy of Laurie Duckham ©

The day has gotten happily away from us—a nourishing distraction from virus-talk and the news cycle. The sublime purple and pink skies are darkening, telling us that dusk is dawning. Time to saunter homeward and rest our busy heads. As we tramp meditatively back to our respective homes we dare to dream. We dream that one day something like CERES, through collective action, vibrates outwards and becomes the beating heart of our city, our collective mood, our urban soul. CERES is like a castle built in the air. Now we have to put the foundations under it.

Notes

1. For more detail on the mythology (to which we are indebted) see, Freya Mathews, 2000. 'CERES: Singing Up the City'. *Philosophy Activism Nature* 1, pp 1–12.
2. Chook is an abbreviation that Australians and New Zealanders use for chicken.
3. Henry Thoreau, *Walden*, in Carl Bode (Ed.), 1982. *The Portable Thoreau*, p 563.
4. Ludwig Wittgenstein, 1961. *Tractatus Logico-Philosophicus*. London: Routledge and Kegan Paul, p 151.
5. Matthews, 'CERES', p 1.
6. Ibid.
7. Ibid.
8. See Samuel Alexander, 2017. *Art Against Empire: Toward an Aesthetics of Degrowth*. Melbourne: Simplicity Institute.
9. Freya, 'CERES', p 6.
10. Samuel Alexander and Jonathan Rutherford (eds.), 2020. *The Simpler Way: Collected Writings of Ted Trainer*. Melbourne: Simplicity Institute.
11. Helena Norberg-Hodge, 2019. *Local Is Our Future: Steps to an Economics of Happiness*. Byron Bay: Local Futures; Helena Norberg-Hodge, 2016 (3rd edn.). *Ancient Futures*. Byron Bay: Local Futures.
12. David Holmgren, 2018. *RetroSuburbia: The Downshifters Guide to a Resilient Future*. Hepburn Springs: Melliodora Publishing.
13. Freya, 'CERES', p 10.
14. Rob Hopkins, 2009. *The Transition Handbook: From Oil Dependency to Local Resilience*. Totnes: Green Books, p 134.

20

Sojourning Through a Quiet City: Envisioning a Prosperous Descent

We have wandered the inner city already in this book, but this morning we return, in the midst of the Coronaverse, to step into this urban heartland again and feel, not so much its new flow, but its still waters. Under lockdown, the city has not stopped but it has slowed to a creep. Mostly empty trams creak by and traffic—both vehicular and pedestrian—is light beyond living memory. A newspaper report tells us that foot traffic in the city is down more than 40%, although today that seems an understatement.

The urban tramps see and feel a different city than they did before, but we also discern new smells and sounds. Perhaps even the *taste* of Melbourne is different. Furthermore, a sixth sense of the uncanny dawns on us, a sense of paradox, as we become aware that our city (and indeed the world) is at once sick and yet also breathing more deeply, with the foot of industrialism lifted, no doubt temporarily, from its throat. If there is a glimmer of hope in this moment, it is that the stench of petroleum lifts quickly. May all wounds heal so fast and not reopen too easily.

Tragic though the pandemic is—and it is not over yet—we need to remember that COVID-19 is a crisis within a broader ecological and humanitarian crisis. In shorthand, it is a crisis of the political economic order and all that depends on it (and endures and suffers from it). It is also a lesson in humility, a reminder that there are many things humanity simply cannot control, try as we might to force nature to bend to our will.

As we sit here on the steps of St Paul's cathedral overlooking the city centre, the relatively quiet and peaceful scene is both alluring and haunting, as if dropping hints that we're in the deceptively timid eye of a hurricane whose embodied energy dwarfs that of a nuclear arsenal. In 1902 William James published his *Varieties of Religious Experience*; here we are in the shadow of this church observing the varieties of *urban experience*; in this case: a city under lockdown. A parishioner calmly walks past us and smiles softly, inducing in the tramps a pang of longing for an absent faith, a faith that things are going to turn out alright in this world or at least the next. We are not lapsed but collapsed Catholics.

Today our agnosticism weighs heavily on us as we reflect on this damaged earthly world. With a nod to Father Marx, there seems to be no time for the 'opium' of religion to distract us from earthly concerns in our darkening Anthropocene, although perhaps that passing parishioner would think otherwise; that in fact we need to give the celestial bodies *heightened* attention in such times. Let us, for now, agree to compromise, and simply say with Lewis Mumford that 'The inner crisis of our civilisation must be resolved if the outer crisis is to be effectively met'.[1] Thoughts to pursue another day. It is too early (or too late) for religion.

Presently, we know that the Barrier Reef is suffering yet another seemingly terminal bleaching event. The last great earthly forests in Siberia and Brazil are being remorselessly hacked away towards some kind of deathly tipping point. Increasingly there is talk of a Sixth Mass Extinction, and of climate tipping points, while we recall the haunting phrase from James Lovelock that the face of Gaia is vanishing. At the same time, ghastly oceans of poverty still surround small islands of unfathomable wealth, betraying the deception of the trickle-down effect. If we were *real* tramps we'd have a hip flask of whisky with us to exorcise these dark thoughts, to warm this wintery mood, but the absence of such peaty medicine is probably for the best, given that it's only just gone 9am.

Courtesy of Helen Lamb ©

The 'COVID erasure' of the many dying planetary and social ecologies speaks a great deal about the current political economy that locks us into the treadmill of the present, refusing all consideration of consequence and legacy. And here we are again, in the Coronaverse, with nature visiting revenge in a manner that commands the attention of the dumb beast of neoliberalism by attacking its innards—free trade, consumer sovereignty, land rent, and… let's call it out, the ability of capitalism to extract surplus value, always an embodied treasure, from disease-threatened populations.

The social, economic, and political trauma caused by the pandemic raises the prospect of an exit from capitalism, about which, to be sure, there has been prolonged speculation and dispute, but which may finally be underway, or is at least prefigured in the current disruptions the virus has invited onto the global stage. When we read that a barrel of energy-dense crude oil was, for a time, 'selling' for negative $37 per barrel, a sense of dread bubbles up as the clarity of upheaval becomes apparent. Everything is upside down. The crises of care that this pandemic manifests—both of people and planet—are not new, but they have been highlighted by it, the contradictions deepening and increasingly resistant to resolution within the existing order. There will be no vaccines for these maladies of accumulation that invite death and renewal, no healing of the sickened beast.

Capitalism has shown time and time again, however, that it is a dextrous creature, able to twist and turn in hope of avoiding any fatal attacks on its legitimacy and longevity. Indeed, it may not be done quite yet, although its condition seems terminal. But the evidence of capitalism's chronic instability and unsustainability is provided in the natural and social sciences (with the dishonourable exception of mainstream economics), and the humanities, all of which document a series of interlocking deadlocks and defaults at rising spatial scales that manifest as planetary crisis.

This consideration coincides with the dawn of the global urban age and rapid, hypertrophic urbanisation in many parts of the world. Certainly the increasing densification of urban life offers a breeding ground for viruses like COVID-19—not itself a knockdown case against urban densification, but a further word of warning against any casual celebration of the 'compact city'. The seemingly paradoxical intersection between these two simultaneous trends—massive system disintegration and vast physical agglomeration—bears thoughtful consideration. For so long we have been sleepwalking, getting ever nearer the cliff face. In this quiet cityscape a knot forms in the gut as we wonder whether the fall is now finally underway.

Still, if there is one thing the coronavirus shows, it is that collective entities—in this historical moment, neoliberal states and their civil societies—really can act as if the house is on fire when we feel it is urgent enough. Therein, however, lies the catch: *when we feel it is urgent enough*. The question that emerges is: If capitalism is feeding off itself, like a snake eating its own tail, what comes next? This is no tired retreat into old conversations about state socialism or the 'third way'. A range of ecological, social, and financial contradictions indicate that, one way or another, coming months and years will see growing pressure on the global capitalist system and the emergence of new political and economic forms and imaginaries.

As crises deepen and intensify, a further descent of some form is underway, with new ecological, technological, and social realities destined to disrupt (are already disrupting) the status quo. The human challenge is to ensure that the post-capitalist era emerges as far as possible through design rather than disaster, acknowledging all the while that self-determination is a luxury not available to everyone, particularly those facing the violence of and on capitalism's new frontiers.

We saunter from the cathedral steps towards the banks of the great river that dissects our city. Does it flow more cleanly today on account of Melbourne's industrial metabolism having slowed? Has less consumer trash found its way into her waters? In the midst of the current pandemic, which is causing so much human suffering, it is clear that shutting down the aviation industry and much of consumer culture is allowing a moment for the planet to take pause from the onslaught of global industrialism.

For so long we have been told that we just cannot produce less, only more; that the type of economic contraction we see today was not possible. And yet here it is, albeit by disaster, not design. As French philosopher Bruno Latour recently commented in the Twittersphere: 'Next time, when ecologists are ridiculed because "the economy cannot be slowed down," they should remember that it can grind to a halt in a matter of weeks worldwide when it is urgent enough'.

Decades of green censure have done little or nothing to reset the path of growth fetishism, extractivism, and consumerism, and yet suddenly, almost overnight, this pandemic has disrupted the status quo, forcing a massive suspension of capitalism. This invites reflection on whether this new crisis could be a prelude, for better or for worse, to a new economic, political, and social imaginary. Global society has entered a chrysalis era, insecure but with latent potential. Nothing is ordained, and it is the task of politics, or collective action more broadly, to choose the next world.

Courtesy of Mark Henson © (http://markhensonart.com/).

If growth is the defining metaphor of our destructive civilisation, might 'degrowth' help unsettle that dead metaphor from its grave and help usher in an ecological civilisation? We feel there is, in fact, no alternative but some form of degrowth, even though hoping for success in this transformation requires a faith of its own, albeit of a different form to our parishioner friend's. Degrowth evokes what Charles Eisenstein calls 'the more beautiful world our hearts know is possible',[2] and without some guiding vision, it is not clear what today's incessant chatter about a 'transition' means. Transition to where?

If Lenin's question—'What is to be done?'—is asked prematurely, without having a sufficiently clear vision of what sustainably and justice actually demand of us, then there is great risk that one's actions, motivated by the best of intentions, are directed in ways that fail to effectively produce any positive effect (or positive affect) and, indeed, may be counterproductive to the cause. Allow us then to further unpack our guiding vision, as we tramp the theoretical contours of the degrowth paradigm.

First we walk along the nexus of ecology, technology, and growth. Degrowth is a movement that sees the goal of limitless economic growth as being dangerously incompatible with a finite planet. From this perspective, the notion of 'green growth'—where it is heralded that economies will grow but in sustainable ways—is a myth.[3] Despite decades of extraordinary technological advance and deep faith in market mechanisms to bring environmental salvation, the so-called greening of capitalism has only produced ever-greater devastation. How long must we wait? With faith in green growth lost, what is needed, we argue, is an equitable degrowth process that downshifts global material and energy demands to sustainable levels. The growth imperatives of capitalism, however, will not accept this, which is why sustainability implies a post-capitalist world.

Having walked this short and clear path to post-capitalism, we turn Left, specifically to the Left's egalitarian instincts. Environmental concerns cannot be isolated from social justice concerns, and the growthists always kick back arguing that the only path to poverty alleviation is via the strategy of GDP growth, on the assumption that 'a rising tide will lift all boats'. Given that a degrowth economy deliberately seeks a non-growing economy—on the assumption that a rising tide will *sink* all boats—poverty alleviation must be achieved more directly, via distribution of wealth and power, both nationally and internationally. In other words (and to change the metaphor) a degrowth economy would seek to eliminate poverty and achieve distributive equity not by baking an ever-larger pie but by slicing it differently. Enough, for everyone, forever.

But it is not enough to fall into the ruts of old-school democratic socialism and assume a tax-and-transfer approach would be sufficient. We need to ramble through new social innovations and practices, even if the sweet fruit that is made into new wine sometimes seems that it is being put into old bottles. For example, the shiny new lens of the 'sharing economy' (at least its more progressive formulations) can highlight how more value can be acquired from the same 'slice' of the economic output. By sharing more between households—facilitated by the internet (new wine) or by traditional community engagement (old bottles)—less energy and resource-intensive production needs to occur to meet society's needs. Indeed, even in a contracting economy (whether that contraction is by design or by crisis), households can still secure the tools and other things they need, provided a deeper culture of sharing emerges to replace acquisitive individualism. This soft-sounding notion of sharing entails a bold reinterpretation of 'efficiency' implicit in the degrowth paradigm: produce less, share more. Communities will manage a contracting economy more easily if they share the resources they have.

Nevertheless, in walking this path, how easy it seems for people to misread the map, even though it would seem to be quite clear enough. People might get lost and ask: Is the current economic downturn that we are observing in the city today what degrowth looks like in reality? First of all, let's be clear. Degrowth means *planned* economic contraction. Nothing about the existing economic shutdown in response to COVID-19 was part of the plan for Australia or the world. Indeed, only a few months ago it was almost unthinkable. The plan was economic growth, and then more economic growth. If you think the current contraction means degrowth, you've misread the map, taken a wrong turn. Retrace your steps.

When an economy contracts involuntarily, we know that as recession or, if it lasts long enough, a depression. Nobody advocates such unplanned economic contraction because it has all sorts of negative social effects, including rising unemployment, stress, and poverty. So degrowth must never be confused with recession. The same direction, as everyone knows, can be travelled by different roads. At most the current crisis could be described as 'protective contraction'—an attempt to protect workers, not out of inherent concern for their humanity, but primarily so that they can get back to work as soon as possible, so that capital can extract surplus value from their productive capacity.

But having taken a wrong turn somewhere, it turns out we are where we are, and that is the best (and only) place to begin. Let us not be like the Irishman who, when asked for directions, advises that if you want to get *there*, you'd best not to start *here*. So true, but here we are! In terms of sustainability, there is a real risk (already unfolding?) that everything bounces back to 'normal' levels of growth and consumption as soon as this pandemic passes. History shows that emissions go down during recessions or depressions but tend to rise again as soon as the growth engine restarts.

The question is whether we can manage this pandemic in a way that stops the virus getting (further) out of control, avoids the 'bounce back' to high impact, carbon-intensive living, while also ensuring all people feel economically secure in a downshifted economy. The radical challenge this raises becomes especially stark as we read the early estimates from climate scientists suggesting that the current economic shutdown—extensive though it is—may 'only' produce an 8% drop in global emissions for 2020.[4] What then would 50% emissions reductions look like, let alone a net-zero emissions economy?

Courtesy of Greg Foyster © (http://gregfoyster.com/).

The deep decarbonisation and degrowth required for such contraction would clearly require significant shifts in the ways our economies and cities are structured, including exploring innovative new ways to govern access to land and housing, and having difficult but compassionate conversations about things like wealth redistribution and population growth. And if the response to COVID-19 shows us anything it is that governments can mobilise extraordinary amounts of money when there is political will. This is good news for funding a transition to renewable energy, for example—if we can develop the political will. But let's not try to replace the full energy services of carbon civilisation; let's be content with energy sufficiency. Fortunately, just enough is plenty.

A degrowth transition would also mean a cultural recognition that high-consumption lifestyles are unsustainable and that only lifestyles of material sufficiency, moderation, and frugality are consistent with social and ecological justice. This challenges us to reimagine the good life beyond consumer culture, thereby laying the seeds for a politics and economics of sufficiency. Social movements will be needed to help create the support for these structural and cultural shifts. These might include post-consumerist movements that are prefiguring degrowth cultures of consumption by embracing material simplicity as a path to freedom, meaning, and reduced ecological burdens; community-led resistance and renewal movements; transgressive and creative forms of the sharing economy as a means of thriving even in a contracting biophysical economy; as well as other social movements and strategies that are seeking to develop new (or renewed) informal economies 'beyond the market'. (For interested readers we have unpacked the degrowth paradigm in much more detail in our book *Degrowth in the Suburbs: A Radical Urban Imaginary*.)

So, while the pandemic continues to unfold, as a society we need to consider whether our ambitions are merely to return to business as usual. Or do we aspire for a radical and final break from neoliberal globalisation—shaken awake by this disruption—and aim instead to transition to a social form that prioritises human wellbeing and ecology over material accumulation?

Down by the river your tramps feel the wind change and the temperature drop. We return to Elizabeth Street and again walk her tarseal

Credit: Jessica Perlstein © (http://dreamstreamart.com/).

northward, haunted by the knowledge that a William's Creek still runs beneath us. As we look around our city in crisis, it feels we've walked this path before.

There is no reason to believe that the current season of forced degrowth represents a permanent and final dislocation of the growth machine ambitions of neoliberalism. The relatively recent experience of the Global Financial Crisis (2008–2009) and its aftermath is a worrying precedent. We fear this replay for the current crisis—our anxieties deepened by the observation earlier that neoliberalism is a particularly historically insentient beast. The forces willing for snap back are immense and omnipresent, throughout the Global North. It's easy to highlight, not to say pillory, the 'let's re-open for business' cant of President Trump, but as Wolfgang Streeck reminds us, the European Union is a deeply neoliberal institution, essentially a free trade bloc, that is equally committed in the current historical moment to the earliest possible resumption of the growth machine. Australia is in the same vein. The centre-Left and green parties typically operate within the same growth paradigm, too often committed to little more than a limp 'third way' that talks of 'greening capitalism' or giving it a human face. But that is merely going down the wrong road more slowly.

But caution is advised. The cloak of pessimism is too often the disguise of determinism, a tendency that we reject as bad science and politics. Both defeats and victories are snapped from the jaws of historical crises and it's far too early now to say what will come from the current *degrowth moment* which we, with the support of Prime Minister Scott Morrison, can type as *lockdown*. We write, in May 2020, in the steaming mists of volcanic eruption of the economy and of everyday life. New (or are some old?) social shadows and shapes are discernible. We see people freed from the neoliberal work frame (often harshly) and finding their way in life under a newly assertive state.

A dialectical play of possibilities is evident and certainly too many to try and list here now. But we cannot fail to see on the one progressive hand the radical reassertion of the state and of its care infrastructure as well as the freeing of households from the treadmill of neoliberal work order (and all the fractured and gendered coping reflexes that went with it). Equally, we discern and recoil from the authoritarian possibilities unleashed by new state arrogation, especially in Anglophone nations where populist conservatives reign. Who knows what will emerge from this historical clash of possibilities? Our bleakest vision is the emergence of authoritarian states that will 'lock down the snap back'—that is, reanimate the Earth-eating monster and drive us harder and faster to the graveside of capitalism.

On better days, we hope-think for transition, however messy it might be, to a different social order that finally accepts new ideas of growth and progress. And what mature human being doesn't desire a life marked by growth and self-realisation, a promise-idea seeded most wondrously by the Enlightenment? The simple point of degrowth, and of most radical thought traditions under capitalism, is that this journey mustn't consume the social and ecological substrates that sustain us.

It may be that ever-deepening crisis in the existing system of capitalism is the most likely spark for a paradigm shift both in the political economy of growth and its cultural underpinnings. To say this, however, is not to romanticise crisis like dreamy-eyed optimists. In fact, our theory of change is based on a deep pessimism about the prospects of smoother and less disruptive modes of societal transformation. As the pandemic deepens or exacerbates the range of pre-existing crises, it seems that our collective task now is to ensure that these destabilised conditions are used to advance progressive humanitarian and ecological ends, rather than exploited to further entrench the austerity politics of neoliberalism.

How to ground this great and terrible opportunity in everyday life? For those who recognise the potential in this moment to think and act differently, our basic function is to keep hopes of a radically different and more humane form of society alive. The encounter with crisis can play an essential consciousness-raising role, if it triggers a desire for and motivation towards learning about the structural underpinnings of the calamity itself.

We believe social movements should be preparing themselves to play that educational role, and in fact it is heartening to see this already unfolding in the many inspiring social responses to this tragic time. Amongst many examples of this, we highlight but one: David Holmgren and the permaculture movement, who are mobilising as we write for the creative renewal of our cities and suburbs. Holmgren's relaunch of his visionary book *RetroSuburbia: The Downshifter's Guide to a Resilient Future*[5] during the pandemic exemplifies this vision and faith in grassroots activity. And importantly, under its warm messaging about restoration of natural ecology and human values lies a serious prosecution of accumulative capitalism.

In the midst of this pandemic, our challenge is to come together and set sail for newer, safer shores, resisting the sirens of destruction that would woo us back to the sinking Atlantis of capitalism. This is not a time of species affirmation; it is the hour of gravest peril. It is also a reopening of human possibility. To liberate human prospect, we must cast down not defend the burning bridges of a dying capitalist order and be brave enough to entertain the possibility of a permanent and planned economics beyond growth; of protective contraction, of degrowth, culminating in some 'steady state'. This pandemic is an ambivalent invitation—even an incitation—for humanity to confront this turning point in the human story with all the creativity, wisdom, and compassion we can muster. In this, we tramps, all of us, are in search of the new human story as written in the coldly glowering text of the industrial city.

Notes

1. Lewis Mumford, 1973. *The Condition of Man*. London: Mariner Books.
2. Charles Eisenstein, 2013. *The More Beautiful World Our Hearts Know is Possible*. Berkeley: North Atlantic Books.
3. See generally, Jason Hickel and Giorgos Kallis, 2019. 'Is Green Growth Possible?' *New Political Economy*. DOI: https://doi.org/10.1080/13563467.2019.1598964.
4. See Josh Gabbatiss, 2020. 'IEA: Coronavirus Impact on CO_2 Emissions Six Times Bigger than 2008 Financial Crisis'. *Carbon Brief* (30 April 2020).
5. David Holmgren, 2018. *RetroSuburbia: The Downshifter's Guide to a Resilient Future*. Hepburn Springs: Melliodora Publishing.

21

Glitter and Doom: Between Naïve Optimism and Despair

Over the last two centuries in the West, we have been telling ourselves that economic growth is the most direct path to prosperity, that the good life implies material affluence, and that technology and 'free markets' will be able to solve most of our social and environmental problems. In recent decades, the most developed capitalist nations have even attempted to impose this rationality on the entire globe, arrogantly declaring the 'end of history'. COVID-19 is a radical reassertion of history against this preposterous neoliberal puffery. It is a painful, indeed tragic reminder that humans and nature are part of a single unending narrative with many surprises to be expected along the way. A world suddenly broken apart by the unforeseen pandemic knows that it is just another moment in the unfolding historical record. We should not speak of the end of history, only of whether a new beginning is near.

Our well tramped home city of Melbourne is an outgrowth and exemplar of this narrow vision of industrial progress. It is a city that entails deep ecological contradictions and socially corrosive inequalities, while at the same time—another contradiction—it is full of vibrancy, beauty, and intrigue. As this book project has sought to demonstrate, through our incomplete and fragmentary perambulations, there exists both life and death in this 'liveable city'; both enchantment and disturbance, with boundaries sometimes blurring.

In the AC period, the city has been steadfastly locked down by its governing State of Victoria, meaning, as we have tried to explain, that old time-space boundaries have evaporated while new ways of daily life have been embarked upon. It's all a great, and to some degree terrible, experiment, with mixed results so far. Many have been freed from neoliberal work regimes, but countless others have been condemned to unemployment, uncertainty, and for a large group of women and children, new rounds of family stress and violence. The old order's dirty, smouldering secrets have been forced to the surface by the crisis. So who really wants the old order back? We, amongst many, sense a general beckoning in the public mood for something new.

So, as each day passes, and each new crisis arrives, the industrial story of urban development becomes less credible, its future less plausible. With disarming clarity, we 'moderns' see, and increasingly *feel*, that the global economy is degrading the ecological foundations of life, threatening a catastrophe that in fact is already well underway. The fact that capitalism also produces abhorrent inequalities of wealth raises the questions: For whom do we destroy the planet? And to what end? We are told to wait for justice, as if in a Kafkaesque novel, but we are not told how long we must wait.

As if this were not enough, the assault of capitalism strikes deeper still, to the core of our being. Walking the streets of our contradictory city we have often sensed that modern urban life is spreading a spiritual malaise, an apathetic sadness of the soul, as ever-more people discover that 'nice things' in a new apartment or McMansion cannot satisfy the universal human craving for meaning. The abundance of stuff, as technology forecaster Paul Saffo argues, has merely produced new scarcities, creating an existential void that stuff simply cannot fill. Relatively privileged though your urban tramps are, we are not immune to the anxieties, dread, and gloom that can set upon the inhabitants of late capitalism. It is difficult to be human. We must all suffer, as fallen creatures, in unique and qualitatively different ways, so let us aim to make necessary suffering meaningful, for there is nothing worse than meaningless suffering.

21 Glitter and Doom: Between Naïve Optimism and Despair

As the dominant culture continues to pursue an uninspired, narrowly materialistic and carbon-intensive conception of progress, it is becoming clearer that we, *homo urbanis*, are suffering under a gross failure of imagination and being governed by mistaken ideas of freedom and wealth. Our cities—our growth machines—so often reflect this uncreative inertia of the urban development process. Clone towns keep being pasted over mainstreet, with the same mega-corporate food chains and outlets, as if trying to homogenise a universe that has the capacity for infinite diversity.

Courtesy of John Holcroft © (http://www.johnholcroft.com/).

Deep down we all know that something is very wrong with this industrial order and its dominant narrative of urban progress—that there must be better, freer, more humane ways to live. But we live in cities that seem to conspire to keep knowledge of such alternatives from us. Bombarded in every direction by cultural and institutional messages about wealth, innovation, and 'green growth', we are being sold a lie: that consumerism in the suburbs is the peak of civilisation and that elsewhere there is no alternative to hypertrophic densification. Over time, as these messages are endlessly repeated and normalised, our imaginations begin to contract and we lose the ability to envision different worlds and different cities; we lose the ability to conceive of alternative paths of urban prosperity and progress.

The future was hurtling towards us; then it arrived; now it is behind us. The new future—after-COVID—isn't what it used to be. And yet, it seems we have not yet found a new story by which to live. We are the urbanites between stories, desperately clinging to yesterday's but uncertain of tomorrow's. Adrift in the urban cosmos, without a narrative in which to lay down new roots, *homo urbanis* marches on, naively attempting to solve the crises of civilisation with the same kind of thinking that caused them.

But then again, perhaps the new words we need are already with us. Perhaps we just need to live them into existence.

Courtesy of Mark Henson © (http://markhensonart.com/).

Buckminster Fuller, an American architect and systems theorist, once said: 'You never change things by fighting the existing reality. To change

21 Glitter and Doom: Between Naïve Optimism and Despair

something, build a new model that makes the existing model obsolete'.[1] While we sympathise with the sentiment, we resist the either/or dichotomy as being too simplistic. We need both resistance *and* renewal in the city, and it is for each of us to find where, in our own lives, there are cracks in capitalism, in the hope of gaining some critical leverage. After all, as poet and songwriter Leonard Cohen advised, 'ring the bells that still can ring /forget your perfect offering /there is a crack in everything /that's how the light gets in'.[2]

This approach to urban transformation essentially expresses the idea that examples are powerful, that examples can send vibrations through the networked mycelium of our cities, creating cultural currents that can turn into subcultures, that sometimes explode into social movements, and that, on very rare occasions, can spark a revolution in consciousness that changes the world and its structures.

In an age when it can sometimes seem as if there is no alternative to the carbon-intensive, consumer way of life, being exposed to real-world examples of new ways of living and being has the potential to expand and radicalise the urban imagination. At such times, when we close our eyes, new, more hopeful futures flicker in and out of existence, forcing us to vote for an alternative future where once we had thought there was no alternative to casual, progressive ecocide. Is there a legitimate space between the breezy glitter of techno-optimistic fantasies and the demobilising sense of impending doom?

We have tried to find and occupy that space in these pages. Tramping through the landscapes, histories, and cultures of Melbourne has opened our eyes. With an awakened perspective more sensitive both to urban beauty and urban violence, one learns (or one hopes to learn) how to live more compassionately, creatively, and transgressively, with a heightened sense of alternative pathways through and from the city *as it as*.

This experience of possibility can be both exhilarating and terrifying, for in those moments when we are able to break through the crust of conventional thinking we see that the city, as it is, is not how it has to be. Faced with living proof that people, here and now, are living in the spirit of resistance and renewal—if only, at first, in a microcosm—the structures and narratives that define the contours of the urban condition can suddenly seem less compelling. In this way, lived examples of alternative modes of being-on-Earth-in-the-city can challenge us to be examples ourselves—can challenge us to live experimentally within the cracks of capitalism in the hope of setting ourselves and each other free. And if not *free*, then at least *freer*. Nothing mobilises a community of people quite like the taste of freedom. Soon enough, in the flow of revolt, new models threaten to make the old model obsolete.

This is the disruptive potential of even small-scale explorations of new ways of living and being in the city. When shaping post-capitalist forms of life, however, one has to start somewhere—within capitalism—with nothing but bold intentions and reckless hope. It follows that the efforts of those who seek to break new paths in the urban landscape and signal new directions must inevitably seem insignificant at first. Pioneers are easily dismissed as utopian dreamers or escapists who lack a sense of political reality. But just as vision without politics is naive, politics without vision is dangerous. We must dream as we shape our urban politics, or else we will never awaken from the existing nightmare of pragmatism without principle. We are all utopians and always have been; only some of us are more ambitious.

Our task, therefore, is to expose and better understand the myths that dominate our destructive mode of existence, and to envision what urban life would be like, or could be like, if we were to liberate ourselves from today's myths and step into new ones. We search for grounded hope between naive optimism and despair. Without vision and defiant positivity, *homo urbanis* might perish.

Nevertheless, our myths today have become so entrenched that they have assumed a false necessity, which is to say, they no longer seem to be myths at all. Rather, the myths of industrial urbanism—which are the myths of limitless growth, technological redemption, hypertrophic densification, and fulfilment through affluence—seem to be a reflection of some 'grand narrative' from which we cannot escape.

But, as we have already suggested, there is a collective rumbling unfolding across the globe today, a collective rumbling of love and rage, and rage and love, in a world newly stressed by pandemic and the grinding failures of the superpowers. Do you hear this great song of human yearning? It is spreading in all directions, which means it is both coming your way and emanating from you. Currently dormant, our repressed hopes are all embers ready to ignite, awaiting a rush of oxygen that will flare our utopian ambitions. Breathe deeply, they say and demand the impossible. Let us stoke the fire of ecological democracy that is burning in our eyes, not because we think we will succeed in producing a just and sustainable city, but because if we do not try, something noble in our hearts and spirits will be lost. So open your mind, gentle reader, for the future of the city is but clay in the hands of enchanted imaginations. We are being called to make things new, so do not look for Atlantis elsewhere. We must find it, or rather (re)create it, beneath our very feet.

As artist-activist John Jordan says: 'When we are asked how we are going to build a new world, our answer is: "We don't know, but let's build it together"'.[3] If we start in that spirit, we are off walking the city on the right foot.

Notes

1. See Steven Sieden, 2012. *A Fuller View: Buckminster Fuller's Vision of Hope and Abundance for All*. Divine Arts, p 358.
2. Leonard Cohen, 'Anthem', from his 1992 album *The Future*.
3. See Rebecca Solnit, 2016 (3rd edn.). *Hope in the Dark: Untold Histories, Wild Possibilities*. Chicago: Haymarket, p 93.

Part IV

Coming Through Slaughter

22

An Urban Politics of Enchantment

As we acknowledged from the outset, this book was predicated upon, and initiated by, a vague but disturbingly pervasive sense of disenchantment in the contemporary urban world. It is easy, we feel, to be disenchanted in the industrial city; perhaps it is even the default experience of *homo urbanis* today. The industrial aesthetic under machinic neoliberal capitalism is typically ugly, uninspired, and yet often overwhelming, a product of narrow 'efficiency thinking' and flawed priorities which privilege capital and concrete expansion over most else. The cogs of the urban development process turn, inexorably leading to the diminishment of natural and human amenity and life-space. By a 'seeming fate', Thoreau once noted, there is 'no time to be anything but a machine'.[1]

All this has damaging if not destructive affective impacts on the urban mood. Traffic oppresses the pedestrian and agitates the commuter, providing a backdrop of sensory tension in a landscape desperately in need of more music and birds. Road rage is surely a spiritual condition—an urban (un)condition—a reflection of culture and context. Our primordial natures suffocate in a hypertrophic built environment, towering above us and out to the horizon and beyond, lacking human scale. In the shadow of billboards, it should come as no surprise that urban neuroses fester.

Courtesy of James Porto © (http://jamesporto.com/).

Life is so busy—yet busy doing what? It seems the modern city-dweller always has to be somewhere other than where they are. There doesn't seem to be time to reflect collectively on fundamental questions: What is the city *for*? What is the economy *for*? And who or what is answering these questions if we, the people, are not? The last of the rainbow lorikeets might fly overhead and we might be too engrossed in the work-and-spend cycle to notice, our faces disfigured by holding our noses too close to the grindstone. Disenchantment, Jane Bennett notes, names 'an unhappy psychological state'.[2]

Beyond the often-heavy existential dimension of urban life, our lives today take place and form in a context of greater ecological and humanitarian tragedy. *Homo urbanis* is recklessly siphoning resources away from overburdened natural environments to feed the 'consumptive cities' of carbon civilisation, future generations be damned. At the same time, pointing the finger at overconsumption risks deflecting attention away from the underlying political economy of growth that knows only one mode of pursuit: the insatiable production of *more*.

Too much is never enough, even or especially for the richest, all the while billions remain in material destitution within a global capitalist system that is structurally unable to care (unless caring happens to coincide with the profit motive). In a one-dimensional world economy governed by market logic, feeding and clothing the poor, or protecting the environment, are not high priorities when greater profits can be made selling luxury goods to the affluent. This is the morality of the market; the efficiency of 'free trade'. But freedom for whom, we might ask? And efficiency to what end?

All the evidence indicates that our civilisation is in the process of colliding with a range of environmental limits to growth.[3] And everyone should be able to understand that what cannot continue, will stop. One way or another, this civilisation is finished.[4] Empire is dying from its fatal addiction to growth for growth's sake, and in a world of almost eight billion people, one dare not contemplate the human cost of collapse. Around the world, societal and ecological steam is building up in a closed system, suggesting that greater disruptions will occur in the foreseeable future. Perhaps the long descent is already underway, catalysed by COVID-19.

In this darkening Anthropocene, readers would be justified in asking: How dare one write a book on urban enchantment at such times? If Bertolt Brecht were still alive, would he not declare that this is a bad time, not just for poetry, but for any talk of enchantment? Does it not betray an elitism or a callous aestheticism to seek enchantment while the world suffers? After all, in a world where ecocide, financial crisis, war, civil unrest, and creeping fascism loom ominously on the horizon like dark clouds gathering for a perfect storm, a turn to enchantment certainly needs justification. Surely this is a situation that demands a radical political engagement in order to dissipate and transcend the various tragedies already taking form. At first instance, then, turning one's mind to enchantment might seem like a petty indulgence or trivial distraction, reserved for the comfortable few who do not have to worry about the problems of the real world.

These probing questions have not gone unasked by your tramps, but ultimately and consistently we have found that this critical line of inquiry gets things precisely wrong. On the one hand, we can appeal to experience and simply acknowledge with Bennett that 'enchantment can coexist with despair'.[5] There may have been times when readers have even sensed this paradoxical mood informing our own critical observations. Certainly, we have not written an enchanted book. Instead, we have written a book revealing moments of disturbing enchantment and enchanting disturbance, where we have found ourselves more awake to the city's latent capacity to become something new, something more humane, and less violent. We are more awake to the suffering of the city and those who inhabit it, and therefore less able to ignore or marginalise that suffering. And, at the same time, we are more aware of the cracks in capitalism into which we can thrust the crowbar of our oppositional energies in the attempt to transcend the status quo and help regenerate our troubled cities and fractured urban identities.

Still, the more powerful response to those who worry about enchantment's dubious ethical credentials is to remind ourselves of the case for enchantment's *ethical necessity*. In line with Bennett, we have defended the position that disenchantment presents an ethical and political danger, insofar as that affective condition can lead to resignation, apathy, and the closing of one's community of care. At its most extreme it can lead to nihilism, narcissism, and even the immobilising weight of dread, all of which depoliticises a citizenry. It is no good developing an incisive and penetrating theory that sheds light on the political economy of sustainability if there are not animated citizens ready and willing to live it into existence. This speaks to the affective foundations of politics, too often ignored.

Bennett fears that 'the acceptance of the disenchantment story, when combined with a sharp sense of the injustice of things by the Left, too often produces an enervating cynicism'.[6] We see this in the world today, and, on reflection, we tramps might sometimes participate in it against our better judgement. For disenchantment, as we have been using the phrase, means inhabiting and experiencing a world that is too often absent of purpose and play, and where the narratives of meaning and hope that once kept the flame of vitality alive have been extinguished by the burdens of modern industrial urban life.

It is precisely when the narrative of disenchantment seems most compelling that an alter-tale so desperately needs to be told. For it is not enough merely to have an intellectual appreciation of our alarmingly distressed moment in history. Reason is a necessary critical tool and evidence should inform all decision-making. So far we are modernists, plain, and simple. But these things are an insufficient source of action and energy. As Enlightenment philosopher David Hume once stated, 'Reason is, and ought only to be slave of the passions, and can never pretend to any other office than to serve and obey them'.[7] More concretely, the point is one can read the news and the latest scientific reports and thereby be 'well informed', but still not act out of disenchantment; not participate in collective action or democratic self-governance for lack of inner energy, wonder, curiosity, and grounded hope.

Therein lies the ethical and political potential of enchantment: it provides the 'affective propulsion' to care and engage; the emotional energy to 'propel ethical generosity'.[8] To ignore the emotional drivers of ethical and political engagement is to ignore a critical part of the human condition, since the affective, the ethical, and the political cannot be cleanly separated. Indeed, Bennett argues that 'if enchantment can foster an ethically laudable generosity of spirit, then the cultivation of an eye for the wonderful becomes something like an academic duty'.[9] Something along those lines led to the production of this book.

Disenchantment creates urban inertia, which needs to be acted upon with some creative force—an affective propulsion—before it can be unsettled, otherwise it just goes on. But, as we have already heard Albert Camus declare, 'one day the "why" arises – and everything begins in that moment of weariness tinged with amazement'. '"Begins"', he says '– this is important'. For us, this new beginning sparked the wanderings described in this book, as we set out to practise the gentle art of urban tramping as method; to walk the city and let the city write us; to be open to what it had to teach. Our point is not that everyone will or should find enchantment in urban tramping. Our point is that enchantment has a role to play in urban politics, so people should seek their own sources of enchantment as part of what it means to be a citizen.

Michel Foucault might have considered urban tramping to be 'technology of the self', an attempt to engage the self by the self, for the purpose of re-fashioning the self into someone new.[10] Perhaps that was partly our goal, and to some extent we have been successful; we are not the same tramps we were when we first set out. But we would go further and suggest that urban tramping can also be deemed a 'technology of the city'—an attempt to engage the city by the self, for the purpose of re-fashioning the city into something new.

On that latter count, our success has been less clear. On the one hand, we find ourselves looking more or less at the same built environment, which should come as no surprise, since tramps step lightly and do not drive bulldozers. On the other hand, in ways we have explained, although we look at the same city, we see and feel different things, due to the moments of urban awakening that have both disturbed and enchanted us. So in that sense the city *has* changed, just as it has changed us. Walking is a simple technology, to be sure, deserving of celebration as much as Ivan Illich celebrated the bicycle.[11] But in words widely attributed to Leonardo da Vinci: 'Simplicity is the ultimate sophistication'.

Courtesy of Greg Foyster © (http://gregfoyster.com/).

Walking is an embodied presence in motion that always begins with a first step, as Rebecca Solnit describes with disarming simplicity:

Muscles tense. One leg a pillar, holding the body upright between the earth and sky. The other a pendulum, swinging from behind. Heel touches down. The whole weight of the body rolls forward onto the ball of the foot. The big toe pushes off, and the delicately balanced weight of the body shifts again. The legs reverse position. It starts with a step and then another step and then another that add up like taps on a drum to a rhythm, the rhythm of walking. The most obvious and the most obscure thing in the world, this walking that wanders so readily into religion, philosophy, landscape, urban policy, anatomy, allegory, and heartbreak.[12]

Thus began our project of urban tramping, a project disrupted midway by COVID-19, a plague that ended up performing our thesis of a fractured modernity vulnerable to its various internal contradictions. As we write (June 2020) this unfolds as a deep epochal crisis, much more important for human prospects than its recent prequel, the Global Financial Crisis of 2007–2008. The GFC was the neoliberal Emperor suddenly unclothed by the inevitable outworkings of financial (and regulatory) chicanery. The State, long derided, came yet again to rescue capitalism from its death drive. After the bailout, things settled rather quickly back to the old disequilibrium of the rattling, spluttering growth machine economy. The COVID pandemic is another matter altogether, a twin crisis of nature (the virus) and society (the neoliberal dispensation). Its deep existential tendrils reach to the heart of things and surely call time on the spluttering beast.

Epochal transformation is not the cracking open of the earthly heavens but the unfolding and compounding of terrestrial crises—eventually a *force majeure* that finally blasts all system defences away and reopens human history to new possibilities. In late May 2020, the pandemic was enjoined by a new global disruption, the death of a black man, George Floyd, in a violent encounter with Minneapolis police. A great disturbance was engendered in the USA with urban masses deploring racial violence (and many other things besides). Protests followed in cities around the globe, including Melbourne (and other Australian cities), where the focus was on the brutality long experienced by Indigenous people. The protest narrative of contemporary racism built on the foundations of colonialism was repeated in different ways in city protests rippling across the world.

It seems that capitalist history, not just the historical moment, is being brought to account. Barack Obama sees 'an incredible opportunity for people to be awakened'.[13] The tramps did not miss his beat. We have, as we told you in the opening of the book, determined to continue the gentle art of tramping in quest for enchantment, in a time of transformation.

Where Have We Been and Where Are We Going?

Books are best understood backwards, even though they must be read forwards. Before closing this book, then, let us briefly lift to the surface and review the central offerings from our urban politics of enchantment, gleaned from our critical perambulations.

Our journeying began as we set out toward the industrial heart of the city, Flinders Street Station, and beneath its great clock we found ourselves temporally dislocated as we reflected on the mind-boggling depth of Australia's human history. It is a sign of a true ecological civilisation for a culture to inhabit a land for tens of thousands of years without fatally degrading it. Conversely, here we are, the uncivilised children of the industrial order, barely three hundred years old, having driven the world to the precipice of planetary Armageddon.

Our modern self-image as *homo sapiens*—meaning 'wise human'—must be deconstructed, for there is nothing wise about the snake that eats its own tail or a cancer that grows itself to death. And just as we must contemplate the deep past to understand our primordial roots and the wisdoms of indigenous cultures that *homo urbanis* must relearn, so too must we look to the *deep future* to understand the responsibility we have to hold this planet in trust for future generations. The horizon of sustainability is not 2050 or even 2100. The sun will burn for billions of years, so let us not destroy the human story on the first page. We must aspire to be pioneers of the deep future,[14] and not simply walk down the wrong road more slowly, which seems to be the extent of ambition in mainstream environmental politics today.

We then scaled the dizzying heights of the Eureka Tower situated in Southgate, and as the sunset and the skies darkened we were left to look out over our 'city of gold' from the 88th floor. It is one thing to know facts about fossil fuel consumption. It is quite another to be struck down by an awesome view of an infernal urban landscape, reaching out to the horizons and beyond, with the distant representation of its glowing energy services at once beautiful and terrifying. It was the thirteenth-century writer Albertus Magnus who once noted that 'wonder is somewhat similar to fear',[15] and that evening up Eureka Tower we embodied that strange sensibility. But we also remembered Bennett's warning that 'fear cannot dominate if enchantment is to be, … [for] overwhelming fear will not becalm and intensify perception but only shut it down'.[16] Accordingly, we turned away from the view so as not to be paralysed by it.

As we descended 300 metres from the viewing platform we came to understand more clearly that a post-industrial economy run entirely on renewable energy will not look much like this city of gold. Beyond carbon civilisation lies an energy descent future, which we maintain implies transcending the growth model of progress and its urban manifestations as the age of energy abundance comes to an end.[17] For now, however, we inhabit an industrial city, one still trying to rebirth itself. That clock at Flinders Street Station is ticking. We heard it especially loudly up Eureka Tower.

From there we wandered through the 'Devil's Playground', a part of Melbourne's inner city that is punctuated by a particularly stark case of hypertrophic vertical sprawl. We have been using the term 'hypertrophic' to denote a contemporary inflation of urban ambition by development capital in many global cities, and acutely in Melbourne, which seeks ever greater human and built densities driven by the lure of profit not human ecology. This is the reality of the compact city ideal of green planners under conditions of freewheeling neoliberal urbanism. The tramps found humans at the place scale resisting the intrusions and extrusions of development capital, at least for now.

All this reaching for the sky by capitalism reminded us of its death drive, emerging from internal contradictions, as Marx explained, that would finally end in system default and transformation. Under neoliberalism, capitalism seemed to the knowing eye to be steadily burying itself in a final historical plot reserved for itself. The inevitable question of death was invoked, leading us to tramp towards its historical (cemetery) and contemporary (towerscapes) urban forms. In these journeys of old and new, we encountered a political economy trapped in a choking hold that must one day prove fatal—a strangulating contradiction; a simultaneous denial and prosecution of death. If only the graves and their grave testaments of human experience could speak to the minds and hearts of contemporary developers who see no limit to urban ambition. Death imposes limits; we must always learn from it.

We then strolled the corridors of consumption at Campbellfield's largest shopping cathedral, experiencing the usual sensory overload and spiritual dis-ease. It was yet another reminder that in every way we still inhabit an industrial city, even if much of the dirtiest manufacturing has been exported to the poorer parts of the world, for us to import superficially 'cleaner' products for our collective consumption.

As we drifted around in a bright and shiny sea of malaise we realised that almost every commodity we could see, including the building itself, had supply-chains and processes of manufacture and distribution that were saturated with fossil fuels, especially oil. And beyond the embodied carbon of these consumer goods, the sheer resource demands of affluent lifestyles make the prospect of globalising this material culture an ecological impossibility. According to the ecological footprint metric, the world would need four or five planets if the global population consumed at levels of the average US or Australian citizen.[18] And yet, despite this context of gross ecological overshoot, even the richest nations and individuals still seek growth without apparent limit, even as the human population continues to rise by around 200,000 souls every day.

A knot forms in our stomach as we ponder how our species will ever get out of this dreadful predicament, but it is loosened slightly as we recall the wise advice from the great twentieth-century British philosopher Bertrand Russell: 'Gloom is a useless emotion'.[19] With defiant positivity we search for a way out of this ghastly shopping mall. The only exit we see is one that involves a collective search for enchantment *beyond consumer culture* and a collective political effort to restructure our

22 An Urban Politics of Enchantment

economies to support enchanted lifestyles of material sufficiency rather than the existing system that so often locks us into the disenchantments of deathly abundance. Soon we will venture to CERES where a more realistic vision of ecological viability is on offer.

Courtesy of John Holcroft © (http://www.johnholcroft.com/).

In a small act of defiance, we not-so-young men decided next to 'go West'. We tramped to Footscray, perched on Melbourne's enduring industrial river the Maribyrnong. The quest was for what the national press had termed 'the last hotel', an old pub holding out against the vicissitudes of neoliberal urban redevelopment. A much more impressive act of defiance than ours, this kind landlord curated a caring place in an uncaring (redevelopment) space, open to lostworlders. In a pub, of all places, we learned about the magic of resistance in the industrial machine.

Our next two excursions brought us face to face with, and indeed part of, resistance and rebellion in the city. The Guardians of Gandolfo Gardens ultimately failed to save the multitude of old-growth trees planted a century prior, but perhaps winning the war is less important than the spirit with which one enters the urban fray. We read about, observed, and participated in this unsuccessful resistance movement and, admittedly, we are still haunted as we walk past the desolate wasteland that the once-flourishing gardens have become. We wonder where the rainbow lorikeets, wattlebirds, and possums will sleep tonight, and where parents with their newborns will find shade from the fierce summer sun. But the guardians did what they could, like Dr Rieux in *The Plague*, not out of heroism but common decency. They fought the 'never-ending defeat', and as Wendell Berry once said: 'We don't have a right to ask whether we're going to succeed or not. The only question we have a right to ask is "what's the right thing to do?"'[20]

The same impulse had us wandering the streets and 'dying' with Extinction Rebellion, not because we think this movement has it all figured out but, again, because it's worth sticking the crowbar into the cracks of capitalism at any and every opportunity in the hope that our modest efforts will somehow contribute to a more tolerable world. As we write we see thousands upon thousands of people on the streets of the US, protesting the George Floyd killing and all it symbolises. Ordinary people, including journalists, are being beaten and shot at by police. It's all messy, painful, and intense. But as David Harvey argues: reclaiming the city 'cannot occur without the creation of a vigorous anti-capitalist movement that focuses on the transformation of daily urban life as its goal'.[21] Whether it is through COVID, George Floyd, Extinction Rebellion, experiments like CERES, or conventional development, urban life is being transformed day by day. And as we reflect on the lockdown under COVID, physical distancing rules impress upon us the immeasurable value of public assembly and collective solidarity that we must never take for granted.

Next we inspected that conscripted army of urban industrialism, the statuary that rather silently inhabits our cities to this day. Silent until your tramps are startled, as we told you, by the sudden intrusion of one of these stony monuments in everyday life. Here we awakened an army of the dead, inspecting, as never really done, the credentials of their claims to history. Often as not, statues tell the keen eye a story of capitalist violence and excess; your tramps recovering hard meaning well beyond the polishing homilies of monumental inscriptions.

As this manuscript goes to production, a statue that commemorated a slave owner in Bristol, England, has been torn down by a crowd of Black Lives Matter activists and cast joyfully into the river, in the hope that it sinks with its racism and never resurfaces. If a statue itself is a symbolic gesture laced with meaning, then so too is throwing certain bronze zombies into the river. The Mayor of Bristol called it 'an iconic moment' and 'felt no sense of loss' as it was torn down.[22] One doesn't need to be a sorcerer with a crystal ball to foresee the drowning of more zombies in the months and years ahead.

Our final excursion before the virus dropped was in and around Upfield, where we pushed a Sisyphean rock around the suburbs, up its dark mountain of carbon-based culture, infrastructure, and underlying political economy of growth, only to watch the rock fall down the other side. We are told by Camus that we must imagine Sisyphus happy and that we should let the struggle in an absurd universe fill our hearts. With trees no longer there to be saved at the Gandolfo Gardens, new struggles must continue elsewhere.

As we trained to and from this once outer suburb (now merely in the middle of the city's reach), we reflected gratefully on the activists who had fought to keep the Upfield train line open, a reminder of how our cities are saturated with untold histories that are responsible for giving form to the urban landscape and shaping those who pass through them. The meaning(s) of suburban life is not 'out there' waiting to be discovered. It must be created.

Enter the Coronaverse

Suddenly this project was pitched into a new place by the advent of COVID-19 in early 2020. Everyone was thrown, including us, by the sudden deconstitution of everything, including neoliberal urbanism. After a short pause, the tramps realised this was just a new (if now deathly, terminal) field of the industrial capitalist order we had set out to examine. With this insight we resumed our tramping under the particularly strong conditions of the lockdown in the State of Victoria, exemplified by our first, overly cautious AC saunter through the pixelated woods of *Walden: the Game*, where the only risk of virus was from dodgy malware. We soon realised, however, that we had no business being in the woods while our thoughts were with and in the city, and while we might agree with Thoreau that 'the walking of which I speak has nothing in it akin to exercise',[23] under the Victorian state law the outward explanation for our tramps thereafter could only be justified in those terms: as exercise. Nevertheless, might not one need to exercise the mind and soul, not merely the legs? Onwards, outwards, and inwards we tramped.

Baudelaire said that each of us must 'Always be a poet, even in prose'.[24] In these darkest times, such injunctions from the wise departed seemed to have more force, so with this in mind we set out looking for poetry for our urban prose. Instincts took us to where this wisdom was perhaps best entombed, the Melbourne cemeteries of our earlier tramping interest. Why here? Well, in a moment of generational peril surely the ancestors have something to tell us about survival. And pretty much the only place to find them in the industrial metropolis is in the graveyards that constitute that other, unrecognised city of the dead. We revisited the Melbourne General Cemetery where free eye association brought much life-focused poetry to our attention. Strange to relate this from a boneyard! We caught text inscribed on the final vaults of human life. Not award-winning stuff but hard-won species wisdom recorded in scraps of funereal verse. And how straightforward was the message of this poetic prose: *Always observe and cherish, love and justice*. Amen.

It was only a short walk from poetry to literary prose, where we strolled the disturbing streets of Oran as represented in Camus' incredibly pertinent novel *The Plague*. Our copy of the book is now defiled with so many pencil underlinings that the point of highlighting specific passages has lost its purpose. A story for our AC times indeed, one of separation and solidarity, as we battle COVID together, alone. In thinking about this remarkable, disturbing, but uplifting text we are drawn into reflections about the therapeutic role of art in troubled times. The term Apocalypse has a dual meaning, not simply referring to the 'end of the world' but also signifying 'a great unveiling or disclosure' of knowledge and understanding. It is likely to be the artist who will contribute most to the human understanding of such a disclosure if, or rather when, the next disruption arrives.

Rather than wallow helplessly as civilisation descends into barbarism, let us hope that our artists, novelists, poets, and filmmakers are up to the task of weaving narratives of human and ecological suffering into a meaningful web of solidarity and compassion—and thereby, perhaps, give birth to a new golden age of Grecian tragedy that offers both an education and cleansing of the emotions and passions. And, as we have argued in this book, to the extent there is an affective foundation to politics, it becomes clear that art also has a political role to play. Herbert Marcuse recognised this long ago, when he wrote: 'The inner logic of the work of art terminates in the emergence of another reason, another sensibility, which defy the rationality and sensibility incorporated in the dominant social institutions'.[25] In other words, art has the enchanting capacity to expand the conditions of possibility by breaking through the petrified social reality and unshackling the human imagination. Far from representing an escape from reality, then, art and the artist can in fact expose the falseness and contingency of the established order, leaving the truth of alternative realities more accessible and perceivable.

The question of care came next to our minds and following closely on its heels the situation of children during the crisis. There is much talk of their suffering during the lockdown and this testimony was not in doubt. Kids isolated from friends and wider family, restricted and restrained—indeed, *locked down*—at the very time their development calls out for free and ranging play and wide social experience. And yet we also witnessed in our city ramblings another important reality, that of adults and children reunited after being freed from the cage of industrial work and schooling regimes. There was everywhere a sense that children were being suffered (to use ancient language) again, meaning that they were seen, heard, and cherished in new ways by their carers. Bad times for the sick, but perhaps care-full times for many (we acknowledge not all) children. This is a COVID affect worth retaining, indeed fighting for.

Our penultimate perambulation took us to the sanctuary of CERES in East Brunswick, where we meandered around a Covidly quiet demonstration project on the first day of winter. Quiet though it was, the energy and vitality of the place was as strong as ever, feeding the soul in mysterious ways. The creek flowed and the birds sang, and if you looked carefully, you might have even seen a seedling push its way up through the rich soil and begin its slow reach for the sun. The scattering of people seemed to be enjoying a moment of regeneration from the pandemic, as we were. You easily forget where you are at CERES, as you walk by a mud hut, a biogas digester, a fugitive chook, and a flourishing organic urban farm. As Freya Matthews writes, at CERES 'life can no longer be contained by our usual scripts'.[26] You believe that other worlds are possible because you are there already, only in microcosm. Your mood shifts in vague but nourishing directions. You trade with the Celestial Empire and leave richer than you came.

We certainly need movements like Extinction Rebellion, Fridays for Future, and, in Australia, the Adani coal mine protests, etc., to fight the existing system and mitigate its most egregious transgressions. If humanity does not mobilise in opposition, capitalism will enthusiastically facilitate our ecological demise even more swiftly. But if capitalism is going to kill itself anyway, we must be careful not to expend too much limited energy towards fighting a dying beast. We also need to build the new city, here and now, within the shell of the old, and it is that prefigurative strategy that makes CERES so inspiring and so necessary. It is a humble, slow, simple place, yet overflowing with viability and vibrancy, but we have come to the realisation that humility, slowness, and simplicity is what prosperity without growth actually means. When the city begins to resemble CERES, our journey into the deep future will have begun.

We feel that none of this will happen at broadscale until two conditions are met. First, the various urban social movements for deep change must be politically joined and mobilised for a great struggle; to ensure

Courtesy of Maria Peña © (http://www.maria-pena.com/).

that the dissolution of industrialism is as peaceful and reconstructive as it can be. This is a project of the masses not of structures nor, heaven forbid, elites. In cities it is a massed linking of the dispersed powers of communities. Second, the State, Chief Butler of capitalism, must be released to new service; a fealty to human and natural flourishing. The challenge will be to link both projects in a new progressive dispensation without losing force of either in the new mix. As we sat on the steps of St Paul's cathedral during our final tramp, in an eerily quiet city, we were faced with the question of what faith means in the Anthropocene. On that question, do not look to us for guidance. We are not priests, we are merely seekers on a pilgrimage, one that is more about an ongoing creative process than an end-state or destination.

Courtesy of Chaz Maviyane-Davies © (http://www.maviyane.com/).

More Day to Dawn

It is time to offer some concluding remarks and bring to an end our journeys together. Our argument has never been that enchantment is *all there is* to ethical and political engagement in the city. Obviously not. We have simply presented it as a valuable and 'positive resource'[27]—as part of the picture—and we are grateful to Jane Bennett for lifting enchantment from the dustbin of modernity and mining it (if you will excuse the industrial metaphor) with such rewarding insight.

Our application of this forgotten value to the urban landscape is, of course, grossly incomplete. We are all too aware of the limitations of our journeys and the fragmented and idiosyncratic nature of our wanderings. They were local and humble, not epic quests, even as we agree with Thoreau that 'every walk is a sort of crusade'.[28] We have not even scratched the surface of our city's potential for urban enchantment (or disturbance), given that each and every street, building, park, or community has enchanting stories to be told that have the potential to jolt us from our default sensory-psychic-intellectual orientations; offering moments that shake us awake and force us see the world afresh with new eyes. There are many more streets in the city than there can be stories told in one book—certainly a book, such as this one, disturbed by a pandemic. And the paths we traced are obviously not the only paths.

But perhaps this fleeting view of our home city of Melbourne is how it should be—thoroughly incomplete—for it is a reminder that the urban tramp always has more tramping to do; a lesson that may never get old. If we can offer one small, practical guide note with which to close this book, it is to encourage readers to walk their own city more, to seek moments of urban awakening in the wonder of minor experiences that lie waiting for those who go sojourning through civilised life. As Lauren Elkin notes, we should endeavour to become individuals 'keenly attuned to the creative potential of the city, and the liberating possibilities of a good walk'.[29] Previous generations beheld the city face to face; we, through their eyes. But why should not urbanites today also enjoy an original relation to the universe? After all, one never steps into the same city twice, and, as we have said, some things can be more easily *experienced* by the freethinking freewalker than *explained*. These are some of the many mysteries enjoyed by the urban tramp.

Perhaps one clear winter's night you will be sauntering through the sleeping suburbs with the sensibility that Thoreau would have embodied as he sauntered through Walden Woods, and you too will come across an owl perched on a white picket fence, almost close enough to touch, consecrated by silver moonlight, and you will stare at each other in wonder and fascination until chronological time dissolves into a moment of pure presence. You will see the intelligence in this creature's eyes, and it might seem to humbly bow its head as you suddenly realise what you always already knew, that you must live a life that supports the co-existence of this city and that sublime creature. And then, as eternity slowly fades, this owl of Minerva will spread its wings and take flight into the night, leaving the urban tramps looking up at the moon, entranced by the unceasing eloquence of silence. Before long the atmosphere will be washed over by the quiet murmur of passing traffic far away, which, late this evening, resembles the sound of distant waves crashing softly on the shores of some new Atlantis.

Let us head off in that direction, we say, toward new urban horizons, knowing that as *homo urbanis* steps closer, those horizons will move further away. If the sun is just a morning star, then that is all the more reason to keep walking. Our cities have more day to dawn, and we should seek that new dawn, even or especially when the hour is darkest.

Notes

1. Henry Thoreau, *Walden*, in Carl Bode (ed.), 1982. *The Portable Thoreau*, p 261.
2. Jane Bennett, 2001. *The Enchantment of Modern Life: Attachments, Crossings, and Ethics*. Princeton: Princeton University Press, p 7.
3. Graham Turner, 2019. 'Is a Sustainable Future Possible?' *Journal & Proceedings of the Royal Society of New South Wales* 152(1), pp 47–65.
4. Rupert Read and Samuel Alexander, 2019. *This Civilisation is Finished: Conversations on the End of Empire—And What Lies Beyond*. Melbourne: Simplicity Institute.
5. Bennett, *Enchantment*, p 159.
6. Ibid., p 13.
7. David Hume, 2018 [1739–40]. *A Treatise on Human Nature*. New York: Createspace, p 146.
8. Bennett, *Enchantment*, p 3.
9. Ibid., p 10.
10. For an excellent discussion of this aspect of Foucault's work, see Edward McGushin, 2007. *Foucault's Askesis: An Introduction to the Philosophical Life*. Evanston: Northwestern University.
11. Ivan Illich, 1973. *Tools for Conviviality*. London: Marion Boyars; Ivan Illich, 1974. *Energy and Equity*. London: Marion Boyars.
12. Rebecca Solnit, 2001. *Wanderlust: A History of Walking*. London: Penguin, p 3.
13. See Delilah Friedler, 2020. 'Obama Sees Hope in Protests: "There is Something Different Here."' *Mother Jones* (3 June 2020).
14. Samuel Alexander and Amanda McLeod (eds.), 2014. *Simple Living in History: Pioneers of the Deep Future*. Melbourne: Simplicity Institute.
15. As quoted in Bennett, *Enchantment*, p 5.
16. Bennett, *Enchantment*, p 5.
17. See Samuel Alexander and Brendan Gleeson, 2019. *Degrowth in the Suburbs: A Radical Urban Imaginary*. Singapore: Palgrave.
18. See Global Footprint Network: https://www.footprintnetwork.org/ (accessed 5 June 2020).

19. Bertrand Russell, 2009. *Russell: The Basic Writings of Bertrand Russell*. London: Routledge, p 45.
20. Wendell Berry, 2013. 'Confronting the Consequences of Runaway Capitalism'. Interview with Bill Moyers, 7 October 2013.
21. David Harvey, 2013. *Rebel Cities: From the Right to the City to Urban Revolution*. London: Verso, p xvi.
22. ABC, 2020. 'Bristol Mayor Says Edward Colston Statue's Destruction an "Iconic Moment" for the City'. *ABC* (7 June 2020).
23. Henry Thoreau, 'Walking', in Carl Bode (ed.), 1982. *The Portable Thoreau*. New York: Penguin, p 596.
24. Charles Baudelaire, 1990 [1887]. *Intimate Journals*. London: Picador, p 57.
25. Herbert Marcuse, 1978. *The Aesthetic Dimension: Toward a Critique of Marxist Aesthetics*. Boston: Beacon Press, p 7.
26. Freya Mathews, 2000. 'CERES: Singing Up the City'. *Philosophy Activism Nature* 1, p 6.
27. Bennett, *Enchantment*, p 15.
28. Henry Thoreau, 'Walking', in Carl Bode (ed.), 1982. *The Portable Thoreau*. New York: Penguin, p 593.
29. Lauren Elkin, 2017. *Flâneuse: Women Walk the City in Paris, New York, Tokyo, Venice, and London*. London: Vintage, pp 22–23.

GPSR Compliance

The European Union's (EU) General Product Safety Regulation (GPSR) is a set of rules that requires consumer products to be safe and our obligations to ensure this.

If you have any concerns about our products, you can contact us on

ProductSafety@springernature.com

In case Publisher is established outside the EU, the EU authorized representative is:

Springer Nature Customer Service Center GmbH
Europaplatz 3
69115 Heidelberg, Germany

www.ingramcontent.com/pod-product-compliance
Ingram Content Group UK Ltd.
Pitfield, Milton Keynes, MK11 3LW, UK
UKHW021255180426
11947UKWH00011B/796